Automotive Trades

INFORMATION PROCESSING SKILLS:

MATHEMATICS

Automotive Trades

INFORMATION PROCESSING SKILLS:

MATHEMATICS

Thomas G. Sticht
Barbara A. McDonald

GOALS
Glencoe Occupational Adult Learning Series

Macmillan/McGraw-Hill

New York, New York   Columbus, Ohio   Mission Hills, California   Peoria, Illinois

This program was prepared with the assistance of
Chestnut Hill Enterprises, Inc.

Automotive Trades Information Processing Skills: Mathematics
Copyright © 1993 by the Glencoe Division of Macmillan/McGraw-Hill
School Publishing Company.

Send all inquiries to:

GLENCOE DIVISION
Macmillan/McGraw-Hill
936 Eastwind Drive
Westerville, OH 43081

ISBN 0-07-061531-4

2 3 4 5 6 7 8 9 0   POH   99 98 97 96 95 94 93 92

Table of Contents

Introduction

As an employee in an auto repair shop, you will need to do a number of mathematics tasks. You'll use math whether you are a technician, a sales person, a service manager, or an assistant service manager. *Mathematics: Information Processing Skills* is designed to teach you how to do these mathematics tasks. Use this book with the *Automotive Trades Knowledge Base*.

❑ Mathematics helps you to handle **financial resources** on the job. One task in the financial resource area is making estimates to tell customers what the work will cost. Another task is making the final bill. A third task is calculating the company payroll.

❑ Mathematics is needed in another area: working with **material resources** on the job. You use mathematics in ordering parts for the shop. You use mathematics in working with tools on the job and in finding out what maintenance is needed on a car.

❑ The third area where mathematics is used is in handling **human resources**. Human resources are people. You use mathematics in producing graphs about your shop's customers.

In Part 1 of this book, you will try some mathematics calculations to see how much you remember about some basic math skills, such as adding, subtracting, multiplying, dividing, and working with fractions and decimals. You will check your answers when you finish Part 1. Then you'll write the results in the table on page 143. It will tell you which skills you should review.

Once you find which math skills you need to work on, turn to Part 7 of this book. Part 7 contains instruction on all the important mathematics skills.

After you have refreshed your mathematics skills in Part 7, you will be ready to proceed through the rest of this book.

In Part 2 of this book, you will learn some mathematics information processing skills that will make mathematics tasks easier for you.

Part 3 shows you how to use these information processing skills as you perform the mathematics tasks in this book. Parts 4, 5, and 6 show you how to do mathematics tasks in the areas of financial resources, material resources, and human resources. You will use information from the *Automotive Trades Knowledge Base* as you work on these tasks.

PERFORMING MATHEMATICS CALCULATIONS

In this part of the book, you will do mathematics calculations. This is a chance for you to try some basic calculations to see how much you remember. If you are having trouble with the calculations, you can refresh your skills by using Part 7 of this book. But first, try to do the following problems:

1. At A&P Auto, you have records of the cars serviced each week of the month. In one month, the number of cars serviced was 65, 101, 79, 23, and 122. How many cars were serviced in all? (Add the numbers.)

2. You work in the tire sales department. You know that there were 1,200 tires in stock at the beginning of the month. There are 825 tires left. How many tires were sold? (Subtract the numbers.)

3. The time sheets for the office show that 8 people each worked about 15 hours on oil changes this month. Your supervisor wants to know about how many hours were spent on this kind of work in all. Multiply to get an estimate.

4. You want to see what kind of mileage a car is getting. The car used 12 gallons of gas to go 264 miles. Divide to find the miles per gallon.

5. You are working on repairing a windshield washing system. You need to see how much plastic tubing you need in all. The lengths of the tubing needed are 18¾ inches and 23⅜ inches. Add the two lengths to find the total length of tubing needed.

6. You are cutting a rod that is 16½ inches long. It should be 8¾ inches. How much do you have to cut from the rod? (Subtract the numbers.)

7. You have a drum of oil that holds 25½ gallons. About ⅓ of the drum is full. About how much oil is in it? (Multiply. Write your answer in lowest terms.)

8. You are ordering transmission fluid. It comes in 10 gallon drums. Each transmission that you fill uses ½ gallon of fluid. How many transmissions can you fill from one drum? (Divide 10 by ½.)

9. You are filling out a bill for a customer. The technician used Battery UB-111 and Oil Filter 12J. Look up the cost of each part in Figure 3-18 of the *Automotive Trades Knowledge Base*. How much does each part cost? What is the total cost? (Add.)

10. You are billing a customer $59.25. To figure the sales tax, multiply by 0.08. What is the sales tax?

11. You are working at the service desk. You give Mrs. Preston the completed bill shown in Figure 2-1 of the *Automotive Trades Knowledge Base*. Your supervisor says that it is all right for Mrs. Preston to pay only part of her bill. She pays you $25 towards labor. Subtract that amount from the balance to find the amount she owes for labor.

12. The office time cards show that a trainee technician put in 7.25 hours per day. Of that time, she spent 0.75 hours learning safety procedures. What portion of time is that? (Divide 0.75 by 7.25. Round your answer to hundredths.)

13. Six cans of primer paint were used for 8 cars. How many cars can you prime with 9 cans? (Write and solve a proportion.)

14. It takes 4 gallons of oil to completely fill the oil pan of a car. You check the car and find that it needs oil. You add 1 gallon of oil. What percentage of the total oil is that? (Write and solve a problem using percents.)

15. Every year A&P pays for service on its computerized diagnostic equipment. Last year the cost for the service was $500. This year the company tells you the cost will be $580. What percent increase is this?

16. Oil filters usually cost $60 a dozen. The office supply store has them on sale for a 20 percent discount. What is the new price?

17. Figure 2-3 in Chapter 2 of the *Automotive Trades Knowledge Base* shows a portion of a company payroll register. Look in the column labeled "Wages" to answer these questions.
    - ❑ What is the mean of the wages paid to these employees?
    - ❑ What is the median wage paid?

18. You plan to put shelves around the perimeter of a rectangular storage room. If the sides of the rectangle are 12 feet and 6 feet, what is the perimeter of the room?

19. The customer service area needs a new tile floor. The room is a square with 11 feet on a side. How many square feet of tiles will be needed to cover the floor?

20. You are asked to figure the volume of the used-oil storage tank. The radius of the tank is 3 feet. The height of the tank is 10 feet. Think of the tank as a cylinder. What is the volume of the tank in cubic feet?

PART 2

OVERVIEW OF THE THREE Cs

❏ THE THREE Cs: COMPREHENSION, COMPUTATION, COMMUNICATION

Probably when you learned mathematics in school, you spent most of your time learning and practicing basic mathematics skills. You added, subtracted, multiplied, and divided with different kinds of numbers. You solved a lot of math problems where someone else gave you the numbers to work with. When you do mathematics in a work setting, however, no one gives you math problems to solve. Instead, you have to:

❏ Decide what task you need to do and how mathematics can help you do it. You need to put together the numbers to work with.

❏ Decide what computation you need to do and then do it. This is where you use your basic math skills.

❏ Communicate to someone else about the results of your work.

Three kinds of mathematics information processing skills are needed to achieve these tasks. These are the "3 Cs":

❏ **Comprehend** what the task is.

❏ **Compute.**

❏ **Communicate** to yourself and to others about the work and the answers.

COMPREHEND

Comprehend is the first of the 3 Cs. In order to do mathematics, you must first **comprehend** the problem. *Comprehend* means "understand." When you understand a problem, you can solve it. You can compare solving a mathematics problem to driving a car. Before you drive your car, you must know where you need to go. Then you can decide on the route to take and drive there. When you do a math problem, you must know first what you want to accomplish. Then you can decide how to do it. Often people see the numbers in a math problem and immediately add, subtract, multiply, or divide them. This is like jumping in the car and starting to drive without thinking about where you want to end up.

To comprehend a math problem, you must read the problem and understand what it says. You can improve your understanding by using a few reading strategies. Read the problem over two or more times. Try saying the problem in your own words. Then:

❑ State what you are to do. To help yourself, think about what question is being asked. What information is missing? What do you need to find?

❑ Decide what steps to take to solve the problem. Think about how you will gather the necessary information. Remember formulas and rules that apply to this kind of situation. It may help you to think about similar problems that you have solved in the past. How did you solve them?

❑ Find the information you need to solve the problem. Look in the problem itself to see what facts are given. Ignore information that you don't need.

❑ Decide what system of measurement you are using. In your math textbooks, you worked mainly on numbers. In the real world, you always work in some system of measurement. You work with numbers that stand for *time, money, length, weight, temperature, area, volume,* or even *people.* These are the things that are meaningful on the job.

Taking these steps will help you comprehend your task. Once you comprehend the task, it will be easier to solve the problem.

COMPUTE

The second *C* in the 3 Cs is **compute**. After you comprehend a problem, you'll need to decide what kind of computation to do. Then you can perform the necessary computation. Figure 2-1 below shows the kinds of choices you make when you decide how to compute.

The Mathematics Knowledge Base

Basic Operations	Types of Numbers	Systems of Measure
Addition	Whole Numbers	U.S. Standard Measures
Subtraction	Fractions	Metric
Multiplication	Decimals	Length
Division	"Mixed"	Area
	Signed (+;-)	Weight
		Volume
		Time
		Temperature
		Money

Relationships Between Numbers	Statistics and Graphs
Percentages	Count Distributions
Ratios	Central Tendency
Proportions	Variability (Range)
	Bar Graph
	Line Graph
	Circle Graph

Figure 2-1

There are only four types of basic operations: addition, subtraction, multiplication, and division. These are shown in the column at the left.

You choose from among these operations. Sometimes, you'll need to do more than one operation. You may need to add the cost of several checks and then subtract the total from the balance in the checking account, for example.

Sometimes instead of performing operations with numbers, you use numbers to make comparisons. Ratios, percentages, and proportions all describe relationships between quantities. They are shown below the basic operations on Figure 2-1. When you want to see what part of an employee's salary goes for taxes and what part is take-home pay, for example, you show the two parts as percentages.

You perform the basic operations or comparisons on one of several types of numbers. These are shown in the center column at the top. Types of numbers include whole numbers, fractions, decimals, mixed numbers (whole numbers and fractions such as $1\frac{1}{2}$ or $35\frac{5}{8}$), and signed numbers (positive and negative numbers shown with + or - signs). You can add, subtract, multiply, or divide with:

whole numbers: $24 \div 6 \qquad 45{,}768 \times 82$

fractions: $\frac{1}{4} \times \frac{3}{7} \qquad \frac{1}{2} + \frac{5}{6}$

decimals: $0.15 \div 0.8 \qquad \frac{12}{567} \times 0.01$

mixed numbers: $1\frac{4}{5} + 6\frac{7}{8} \qquad 66\frac{1}{2} \times 8$

signed numbers: $^-8 + {^+12} \qquad {^+45} \times {^-19}$

The numbers you work with stand for units in a system of measurement. You must always decide what system of measurement to use. Some systems of measurement are shown in the column at the right on Figure 2-1. You might, for example, do the basic operation of addition on fractions that stand for time. Your answer will be a length of time.

Hours worked: $1\frac{1}{4} \qquad 6\frac{1}{2} \qquad 5\frac{3}{4}$

**System of
Measurement**
↓

Total hours = $1\frac{1}{4} + 6\frac{1}{2} + 5\frac{3}{4} = 13\frac{1}{2}$ **hours**

Another part of the mathematics knowledge base is statistics. Statistics are facts in number form. You are probably familiar with the common statistic called the "mean" or "average." There are other statistics as well. Some of them are shown in the column at the bottom right of Figure 2-1. Number facts can also be presented in graphs. Some common graphs are named on the chart. You may choose to use statistics or graphs to solve a problem.

Once you have decided what computation to do, you do the work. If the computation involves several steps and several operations, you need to be sure to do them all.

COMMUNICATE

The third *C* in the 3 Cs is **communicate**. Communication is very important in any job situation. It's just as important to communicate about mathematics tasks on the job as it is to communicate about things you have read. When you solve a math problem on the job, you do it because someone needs the information. You solve a problem about a payroll so that someone knows what numbers to put on a paycheck. You solve a problem about the cost of parts so that your supervisor can write a check for the parts. You find the estimated cost of repairs so that a customer can decide whether to have the work done.

Sometimes you are the person who needs the information. You find the size of an engine bore so that you can repair the engine.

Usually you communicate the results of your math work in writing. It's very easy to forget numbers unless they are written down. So you write the results in a note to yourself, your supervisor, or a customer. Or you write the results in a log, a register, or a chart. You might prepare a graph to show your results.

❏ HOW THE 3 CS WILL HELP YOU

Using the 3 Cs to solve math problems will help you to get the correct solutions to the problems. It will help you to apply the math skills you have to real-world math problems on the job. You'll have an orderly approach to use each time you encounter a math problem. You may find that you understand what you are doing in mathematics for the first time.

Figure 2-2 summarizes the 3 Cs.

---

1 — Comprehend
  ❏ State what you are to do.
  ❏ Decide on the steps you should follow.
  ❏ Collect the necessary information.
  ❏ Decide what system of measurement to use.

2 — Compute
  ❏ Decide what computation to do. This involves deciding on the operation to do, the types of numbers you need to use, the system of measurement, and whether to use statistics or graphs as part of the process.
  ❏ Do the computation.

3 — Communicate
  ❏ Communicate the results to yourself and others in writing.

Figure 2-2

HOW TO USE THIS BOOK

This book is designed to help you to do mathematics on the job. It is divided into three kinds of mathematics—mathematics for financial resources, mathematics for material resources, and mathematics for human resources. You will learn about each of these three types of job mathematics using the 3 Cs.

All the mathematics in this book is taught in a way that will be useful for your career in auto technology. Look ahead now at Parts 4, 5, and 6 of this book. You'll see that each contains several job situations. You'll learn about the kind of mathematics that you'll need in these job situations. For example, you'll learn about the mathematics involved in estimating the cost of repairs. This will be useful to you if you have a job at the service desk. You'll learn about reading a micrometer. This will be useful to you if you have a job in the engine repair area.

Here's how to use Parts 4, 5, and 6.

❏ First read about the job situation. This section will tell you something about one of the jobs in an auto repair shop and about some of the work you would do in that job. This section may tell you that some information is found in the *Automotive Trades Knowledge Base*.

For example, the section may tell you to look at a figure in the *Knowledge Base*. Look at the *Knowledge Base* when you are instructed to do so.

The job situation will not ask you to do any mathematics. It will give you an idea of how mathematics is used in this job situation. For example, look ahead to the start of Part 4. The first situation is "Payroll." The text there tells you about how you work with a payroll register.

❏ Next you will see the heading "Task 1." This section will describe one kind of mathematics task called for by the job situation. Read this section carefully. Pay special attention to any formulas or rules that are given.

Look ahead to the start of Part 4. On page 10, you will see that the first task in this job situation is "Task 1—Figuring Gross Pay." This section tells you about figuring an employee's pay before taxes. It gives you the formula for finding the gross pay.

❏ Under each task, you will find a section called "Using the 3 Cs." This section shows you how to solve a math problem called for by the job task. The section gives you a sample problem.

Peggy Laskowitz's time card for last week shows the following hours:

Monday	8 hours
Tuesday	8 hours
Wednesday	8 hours
Thursday	8 hours
Friday	8 hours

Her hourly wage is shown on the payroll register below. Calculate her gross pay for the week. Write the total hours and gross pay in the payroll register.

Below the math problem, you will see the headings "Comprehend," "Compute," and "Communicate." These are the 3 Cs. Under each heading is a set of steps and the answer you would write if you were solving the problem.

Read the problem and read each step. Try to think of the answer you would give before you read the answer in the book.

❏ After the example, you'll see the heading "Practice." Following this heading are a number of math problems for you to do. They will be like the one worked out in "Using the 3 Cs." Read each problem, and write your answer for each step of the 3 Cs. Look back at the example in "Using the 3 Cs" for help on completing each step. You will need to refer to information in the *Automotive Trades Knowledge Base* to complete some tasks.

Answers to each problem are found in the Answer Key at the back of the book. Check your progress by comparing your answers to the answers there. You may not have used the same words shown in the answers. This may be all right, if your answer has the same meaning. Check with your instructor.

❏ After you complete all the math problems for the first task, you will go on to do other tasks for the same job situation. Each one will have information on the task, an example under "Using the 3 Cs," and a set of "Practice" exercises.

Check with your instructor to see when you should read the *Knowledge Base* and when you should begin work on Parts 4, 5, and 6.

PART 4

MATHEMATICS AND FINANCIAL RESOURCES

Part 4 covers some financial resources you may deal with on the job. There are two job situations in this part of the book: payroll and estimates/billing. Both situations have to do with money. Each job situation requires you to do several math tasks. Use the 3 Cs to complete each math task.

❏ JOB SITUATION 1 PAYROLL

You work in the business office at A&P Auto Center with Angela, the bookkeeper. One of your jobs is writing paychecks for each employee. You complete a **payroll register** like the one shown in Figure 2-3 of the *Knowledge Base* to find out how much each employee should be paid. Then you write **paychecks** for each employee after you have calculated all deductions.

Employees are paid every week at A&P. The company pays each employee an **hourly wage**. An hourly wage is an amount of money paid for each hour worked. So you need to know how many hours each employee worked each week. You get that information from the employee's **time card**. Employees turn in their time cards each Friday. It's your job to add up the time each employee worked.

You need to know each employee's hourly wage. That information doesn't change from week to week. It is recorded on the payroll register. When the employee gets a raise, you change the hourly wage on the payroll register.

You use the information on number of hours worked and hourly wage to find the employee's **gross pay**. This is the total amount that the employee has earned for the week.

Then you need to calculate **deductions** for taxes. A&P Auto Center has to subtract certain taxes from an employee's pay. These deductions include social security tax and federal income tax. They can include state income tax and other deductions such as disability insurance and health insurance.

The first deduction you make is for **FICA—Federal Insurance Contributions Act**. This is the full name of what people call "social security tax." It is for the employee's social security account, which gives the employee retirement and disability benefits.

A second deduction for each employee is **federal withholding tax (FWT)**. This is a part of the employee's pay set aside for income tax. A&P Auto Center deducts this money from each employee's paycheck. The amount of federal income tax depends on how much money a person makes. Tax also depends on whether the person is married or single. Taxes are lower for a married person. Taxes are also lower when people have **dependents**—people who depend on them for support.

Each month you send the money you withhold for both FICA and FWT to the federal government. It's important to calculate the deductions correctly and to keep accurate records. At the end of the year, the amount of tax you deducted is written on the employee's W-2 form. The employee puts that figure on his or her income tax form.

The amount left over after deductions are taken out is the employee's **take-home** or **net pay**.

All the information on pay and deductions goes in the payroll register. Thursday, after you finish the payroll register, you write a paycheck for each employee. The amount on the paycheck is the net pay. You give the paychecks, along with the payroll register, to Ms. Perez or Mr. Aldo. They sign the checks. Friday morning, you give the paychecks to the employees. The employees can deposit their checks in their banks on Friday at lunch and get cash for the weekend.

TASK 1   FIGURING GROSS PAY

As the assistant to the bookkeeper, you write paychecks for the employees on the payroll register below.

Your first task is to find the gross pay of these employees. Follow these steps:

1. Find the total number of hours worked.
2. Find gross pay, using this formula:

$$\text{hours} \times \text{hourly wage} = \text{gross pay}$$

3. Record the gross pay in the payroll register.

USING THE 3 CS
Peggy Laskowitz's time card for last week shows the following hours:

Monday	8 hours
Tuesday	8 hours
Wednesday	8 hours
Thursday	8 hours
Friday	8 hours

Her hourly wage is shown on the payroll register below. Calculate her gross pay for the week. Write the total hours and gross pay in the payroll register.

COMPREHEND
In the space below, write what you are to do.

> *Find Peggy Laskowitz's gross pay. Write the total hours and gross pay in the payroll register.*

Write the steps you should follow to solve the problem.

> *1. Find the total number of hours worked.*
> *2. Find and record the gross pay.*

Write the information you need to solve the problem.

Hourly Wage: $8.50 per hour.
Hours Worked:

Monday	*8 hours*
Tuesday	*8 hours*
Wednesday	*8 hours*
Thursday	*8 hours*
Friday	*8 hours*

Write the system of measurement.

Time and money

COMPUTE
Do the computation required to complete each task.
Use the steps listed above.

1. *Total number of hours:*
$$8 + 8 + 8 + 8 + 8 = 40$$
2. *hours × hourly wage = gross pay*
$$40 \times \$8.50 = \$340$$

COMMUNICATE
Enter the information in the payroll register.
Enter the total hours and gross pay.

PAYROLL REGISTER   Week Ending 1/10/9X

EMPLOYEE INFORMATION			GROSS EARNINGS			DEDUCTIONS		
NAME	MAR STAT	ALLOW	TOTAL HOURS	REG RATE	GROSS PAY	FICA	FWT	NET PAY
Peggy Laskowitz	M	4	40	8.50	340.00			
Angela Upton	S	1		7.75				
Patrick O'Leary	S	1		10.70				
Jamal Cheston	M	5		11.00				
Barbara Palmer	S	2		8.25				
Sam Li	M	3		9.20				

PRACTICE

1 The time card shows that last week, Angela Upton worked the following hours:

Monday	8 hours
Tuesday	8 hours
Wednesday	8 hours
Thursday	8 hours
Friday	8 hours

Her hourly rate is shown on the payroll register. Calculate her gross pay for the week. Write the total hours and gross pay in the payroll register above.

COMPREHEND
In the space below, write what you are to do.

_____

_____

Write the steps you should follow to solve the problem.

_____

_____

Write the information you need to solve the problem.

_____

_____

_____

_____

_____

_____

Write the system of measurement.

_____

COMPUTE
Do the computation required to complete each task.
Use the steps listed on page 10.

_____

_____

_____

COMMUNICATE
Enter the information in the payroll register on page 11.

2 The time card shows that last week, Patrick O'Leary worked the following hours:

Monday	9 hours
Tuesday	7 hours
Wednesday	8 hours
Thursday	9 hours
Friday	6 hours

His hourly rate is shown on the payroll register. Calculate his gross pay for the week. Write the total hours and gross pay in the payroll register above.

COMPREHEND
In the space below, write what you are to do.

_____

_____

Write the steps you should follow to solve the problem.

_____

_____

Write the information you need to solve the problem.

_____

_____

_____

_____

_____

_____

Write the system of measurement.

_____

COMPUTE
Do the computation required to complete each task.
Use the steps listed on page 10.

_____

_____

_____

COMMUNICATE
Enter the information in the payroll register on page 11.

3 The time card shows that last week, Jamal Cheston worked the following hours:

Monday	5 hours
Tuesday	4 hours
Wednesday	5 hours
Thursday	6 hours
Friday	5 hours

His hourly rate is shown on the payroll register on page 11. Calculate his gross pay for the week. Write the total hours and gross pay in the payroll register.

COMPREHEND
In the space below, write what you are to do.

_____

_____

Write the steps you should follow to solve the problem.

_____

_____

Write the information you need to solve the problem.

_____

_____

_____

_____

_____

_____

_____

Write the system of measurement.

_____

COMPUTE
Do the computation required to complete each task.
Use the steps listed on page 10.

_____

_____

_____

COMMUNICATE
Enter the information in the payroll register on page 11.

4 | The time card shows that last week, Barbara Palmer worked the following hours:

Monday	8 hours
Tuesday	9 hours
Wednesday	9 hours
Thursday	9 hours
Friday	8 hours

Her hourly rate is shown on the payroll register on page 11. Calculate her gross pay for the week. Write the total hours and gross pay in the payroll register.

COMPREHEND
In the space below, write what you are to do.

_____

_____

Write the steps you should follow to solve the problem.

_____

_____

Write the information you need to solve the problem.

_____

_____

_____

_____

_____

_____

Write the system of measurement.

_____

COMPUTE
Do the computation required to complete each task.
Use the steps listed on page 10.

_____

_____

_____

COMMUNICATE
Enter the information in the payroll register on page 11.

5 The time card shows that last week, Sam Li worked the following hours:

Monday	7 hours
Tuesday	8 hours
Wednesday	9 hours
Thursday	7 1/2 hours
Friday	8 hours

His hourly rate is shown on the payroll register on page 11. Calculate his gross pay for the week. Write the total hours and gross pay in the payroll register.

COMPREHEND
In the space below, write what you are to do.

_____

_____

Write the steps you should follow to solve the problem.

_____

_____

Write the information you need to solve the problem.

_____

_____

_____

_____

_____

_____

Write the system of measurement.

_____

_____

COMPUTE
Do the computation required to complete each task.
Use the steps listed on page 10.

_____

_____

_____

COMMUNICATE
Enter the information in the payroll register on page 11.

TASK 2   PAYROLL DEDUCTIONS—FICA

After you find the gross pay for each employee, you find the amount of deductions. You record the deductions for each employee on the payroll register. You subtract the deductions before you write the paycheck.

First, find FICA tax for each employee. The FICA tax rate changes every year or so. For our practice examples, assume the rate is 7.65 percent.

To find the amount of FICA tax, multiply the employee's gross pay by 7.65 percent and round to hundredths.

$$FICA = \text{gross pay} \times 7.65\%$$

USING THE 3 CS

Find the amount of FICA tax to deduct from Peggy Laskowitz's gross pay. Record it in the payroll register below.

COMPREHEND

In the space below, write what you are to do.

Find the FICA tax for Peggy Laskowitz.

Write the steps you should follow to solve the problem.

Multiply gross pay by 7.65%

Write the information you need to solve the problem.

Gross pay from the payroll register is $340.

Write the system of measurement.

Money

COMPUTE

Do the computation required to complete each task.
Use the formula given above.

$FICA = \text{gross pay} \times 7.65\%$

First change 7.65% to a decimal number so that you can multiply.

$$7.65\% = 0.0765$$
$$\$340.00 \times 0.0765 = \$26.01$$

COMMUNICATE

Enter the information in the payroll register below.

PAYROLL REGISTER   Week Ending 1/10/9X

EMPLOYEE INFORMATION			GROSS EARNINGS			DEDUCTIONS		
NAME	MAR STAT	ALLOW	TOTAL HOURS	REG RATE	GROSS PAY	FICA	FWT	NET PAY
Peggy Laskowitz	M	4	40	8.50	340.00	26.01		
Angela Upton	S	1	40	7.75	310.00			
Patrick O'Leary	S	1	39	10.70	417.30			
Jamal Cheston	M	5	25	11.00	275.00			
Barbara Palmer	S	2	43	8.25	354.75			
Sam Li	M	3	39½	9.20	363.40			

PRACTICE

6 Find the FICA tax for Angela Upton. Record it in the payroll register.

COMPREHEND
In the space below, write what you are to do.

_____

Write the steps you should follow to solve the problem.

_____

Write the information you need to solve the problem.

_____

Write the system of measurement.

_____

COMPUTE
Do the computation required to complete each task.
Use the formula given above.

_____

COMMUNICATE
Enter the information in the payroll register on page 17.

7 Find the FICA tax for Patrick O'Leary. Record it in the payroll register.

COMPREHEND
In the space below, write what you are to do.

_____

Write the steps you should follow to solve the problem.

_____

Write the information you need to solve the problem.

_____

Write the system of measurement.

_____

COMPUTE
Do the computation required to complete each task.
Use the formula given above.

_____

COMMUNICATE
Enter the information in the payroll register on page 17.

8 Find the FICA tax for Jamal Cheston. Record it in the payroll register.

COMPREHEND
In the space below, write what you are to do.

_____

Write the steps you should follow to solve the problem.

_____

Write the information you need to solve the problem.

_____

Write the system of measurement.

_____

COMPUTE
Do the computation required to complete each task.
Use the formula given above.

_____

COMMUNICATE
Enter the information in the payroll register on page 17.

9 Find the FICA tax for Barbara Palmer. Record it in the payroll register.

COMPREHEND
In the space below, write what you are to do.

_____

Write the steps you should follow to solve the problem.

_____

Write the information you need to solve the problem.

_____

Write the system of measurement.

_____

COMPUTE
Do the computation required to complete each task.
Use the formula given above.

_____

COMMUNICATE
Enter the information in the payroll register on page 17.

10 Find the FICA tax for Sam Li. Record it in the payroll register.

COMPREHEND
In the space below, write what you are to do.

_____

Write the steps you should follow to solve the problem.

_____

Write the information you need to solve the problem.

_____

Write the system of measurement.

_____

COMPUTE
Do the computation required to complete each task.
Use the formula given above.

_____

COMMUNICATE
Enter the information in the payroll register on page 17.

TASK 3   FEDERAL TAX DEDUCTIONS

When new employees start work, they fill out a form telling whether they are married. They put on the form the number of **withholding allowances** they have. Employees can take one withholding allowance for themselves and one for each dependent. You write this information on the payroll register.

❏ The column headed "MAR STAT" tells whether the person is married (M) or single (S).
❏ The column headed "ALLOW" tells how many withholding allowances the person has.

As the payroll clerk, you use a government booklet called _Circular E_ to figure federal tax deductions (FWT). Sample pages from this booklet are shown on pages 21 and 22. The booklet has tax tables for married and single people and for weekly pay periods, monthly pay periods, and so on.
      To figure an employee's federal withholding tax and net pay:

1. Find the row for the employee's gross pay in the married or single tax table in _Circular E_. Find the column for the number of allowances for the employee. Look in that column for the tax to withhold.
2. Record that number in the payroll register.

SINGLE Persons–WEEKLY Payroll Period

(For Wages Paid After December 1990)

And the wages are–		And the number of withholding allowances claimed is–										
At least	But less than	0	1	2	3	4	5	6	7	8	9	10
		The amount of income tax to be withheld shall be–										
$0	$25	$0	$0	$0	$0	$0	$0	$0	$0	$0	$0	$0
25	30	1	0	0	0	0	0	0	0	0	0	0
30	35	1	0	0	0	0	0	0	0	0	0	0
35	40	2	0	0	0	0	0	0	0	0	0	0
40	45	3	0	0	0	0	0	0	0	0	0	0
45	50	4	0	0	0	0	0	0	0	0	0	0
50	55	4	0	0	0	0	0	0	0	0	0	0
55	60	5	0	0	0	0	0	0	0	0	0	0
60	65	6	0	0	0	0	0	0	0	0	0	0
65	70	7	0	0	0	0	0	0	0	0	0	0
70	75	7	1	0	0	0	0	0	0	0	0	0
75	80	8	2	0	0	0	0	0	0	0	0	0
80	85	9	3	0	0	0	0	0	0	0	0	0
85	90	10	3	0	0	0	0	0	0	0	0	0
90	95	10	4	0	0	0	0	0	0	0	0	0
95	100	11	5	0	0	0	0	0	0	0	0	0
100	105	12	6	0	0	0	0	0	0	0	0	0
105	110	13	6	0	0	0	0	0	0	0	0	0
110	115	13	7	1	0	0	0	0	0	0	0	0
115	120	14	8	2	0	0	0	0	0	0	0	0
120	125	15	9	2	0	0	0	0	0	0	0	0
125	130	16	9	3	0	0	0	0	0	0	0	0
130	135	16	10	4	0	0	0	0	0	0	0	0
135	140	17	11	5	0	0	0	0	0	0	0	0
140	145	18	12	5	0	0	0	0	0	0	0	0
145	150	19	12	6	0	0	0	0	0	0	0	0
150	155	19	13	7	1	0	0	0	0	0	0	0
155	160	20	14	8	1	0	0	0	0	0	0	0
160	165	21	15	8	2	0	0	0	0	0	0	0
165	170	22	15	9	3	0	0	0	0	0	0	0
170	175	22	16	10	4	0	0	0	0	0	0	0
175	180	23	17	11	4	0	0	0	0	0	0	0
180	185	24	18	11	5	0	0	0	0	0	0	0
185	190	25	18	12	6	0	0	0	0	0	0	0
190	195	25	19	13	7	0	0	0	0	0	0	0
195	200	26	20	14	7	1	0	0	0	0	0	0
200	210	27	21	15	9	2	0	0	0	0	0	0
210	220	29	22	16	10	4	0	0	0	0	0	0
220	230	30	24	18	12	5	0	0	0	0	0	0
230	240	32	25	19	13	7	1	0	0	0	0	0
240	250	33	27	21	15	8	2	0	0	0	0	0
250	260	35	28	22	16	10	4	0	0	0	0	0
260	270	36	30	24	18	11	5	0	0	0	0	0
270	280	38	31	25	19	13	7	0	0	0	0	0
280	290	39	33	27	21	14	8	2	0	0	0	0
290	300	41	34	28	22	16	10	3	0	0	0	0
300	310	42	36	30	24	17	11	5	0	0	0	0
310	320	44	37	31	25	19	13	6	0	0	0	0
320	330	45	39	33	27	20	14	8	2	0	0	0
330	340	47	40	34	28	22	16	9	3	0	0	0
340	350	48	42	36	30	23	17	11	5	0	0	0
350	360	50	43	37	31	25	19	12	6	0	0	0
360	370	51	45	39	33	26	20	14	8	2	0	0
370	380	53	46	40	34	28	22	15	9	3	0	0
380	390	54	48	42	36	29	23	17	11	5	0	0
390	400	56	49	43	37	31	25	18	12	6	0	0
400	410	57	51	45	39	32	26	20	14	8	1	0
410	420	59	52	46	40	34	28	21	15	9	3	0
420	430	61	54	48	42	35	29	23	17	11	4	0
430	440	64	55	49	43	37	31	24	18	12	6	0
440	450	67	57	51	45	38	32	26	20	14	7	1
450	460	70	58	52	46	40	34	27	21	15	9	3
460	470	73	61	54	48	41	35	29	23	17	10	4
470	480	75	64	55	49	43	37	30	24	18	12	6
480	490	78	67	57	51	44	38	32	26	20	13	7
490	500	81	69	58	52	46	40	33	27	21	15	9
500	510	84	72	61	54	47	41	35	29	23	16	10
510	520	87	75	63	55	49	43	36	30	24	18	12
520	530	89	78	66	57	50	44	38	32	26	19	13
530	540	92	81	69	58	52	46	39	33	27	21	15

MARRIED Persons–WEEKLY Payroll Period
(For Wages Paid After December 1990)

And the wages are–		And the number of withholding allowances claimed is–										
At least	But less than	0	1	2	3	4	5	6	7	8	9	10
		The amount of income tax to be withheld shall be–										
$0	$70	$0	$0	$0	$0	$0	$0	$0	$0	$0	$0	$0
70	75	1	0	0	0	0	0	0	0	0	0	0
75	80	1	0	0	0	0	0	0	0	0	0	0
80	85	2	0	0	0	0	0	0	0	0	0	0
85	90	3	0	0	0	0	0	0	0	0	0	0
90	95	4	0	0	0	0	0	0	0	0	0	0
95	100	4	0	0	0	0	0	0	0	0	0	0
100	105	5	0	0	0	0	0	0	0	0	0	0
105	110	6	0	0	0	0	0	0	0	0	0	0
110	115	7	0	0	0	0	0	0	0	0	0	0
115	120	7	1	0	0	0	0	0	0	0	0	0
120	125	8	2	0	0	0	0	0	0	0	0	0
125	130	9	3	0	0	0	0	0	0	0	0	0
130	135	10	3	0	0	0	0	0	0	0	0	0
135	140	10	4	0	0	0	0	0	0	0	0	0
140	145	11	5	0	0	0	0	0	0	0	0	0
145	150	12	6	0	0	0	0	0	0	0	0	0
150	155	13	6	0	0	0	0	0	0	0	0	0
155	160	13	7	1	0	0	0	0	0	0	0	0
160	165	14	8	2	0	0	0	0	0	0	0	0
165	170	15	9	2	0	0	0	0	0	0	0	0
170	175	16	9	3	0	0	0	0	0	0	0	0
175	180	16	10	4	0	0	0	0	0	0	0	0
180	185	17	11	5	0	0	0	0	0	0	0	0
185	190	18	12	5	0	0	0	0	0	0	0	0
190	195	19	12	6	0	0	0	0	0	0	0	0
195	200	19	13	7	1	0	0	0	0	0	0	0
200	210	21	14	8	2	0	0	0	0	0	0	0
210	220	22	16	10	3	0	0	0	0	0	0	0
220	230	24	17	11	5	0	0	0	0	0	0	0
230	240	25	19	13	6	0	0	0	0	0	0	0
240	250	27	20	14	8	2	0	0	0	0	0	0
250	260	28	22	16	9	3	0	0	0	0	0	0
260	270	30	23	17	11	5	0	0	0	0	0	0
270	280	31	25	19	12	6	0	0	0	0	0	0
280	290	33	26	20	14	8	2	0	0	0	0	0
290	300	34	28	22	15	9	3	0	0	0	0	0
300	310	36	29	23	17	11	5	0	0	0	0	0
310	320	37	31	25	18	12	6	0	0	0	0	0
320	330	39	32	26	20	14	8	1	0	0	0	0
330	340	40	34	28	21	15	9	3	0	0	0	0
340	350	42	35	29	23	17	11	4	0	0	0	0
350	360	43	37	31	24	18	12	6	0	0	0	0
360	370	45	38	32	26	20	14	7	1	0	0	0
370	380	46	40	34	27	21	15	9	3	0	0	0
380	390	48	41	35	29	23	17	10	4	0	0	0
390	400	49	43	37	30	24	18	12	6	0	0	0
400	410	51	44	38	32	26	20	13	7	1	0	0
410	420	52	46	40	33	27	21	15	9	2	0	0
420	430	54	47	41	35	29	23	16	10	4	0	0
430	440	55	49	43	36	30	24	18	12	5	0	0
440	450	57	50	44	38	32	26	19	13	7	1	0
450	460	58	52	46	39	33	27	21	15	8	2	0
460	470	60	53	47	41	35	29	22	16	10	4	0
470	480	61	55	49	42	36	30	24	18	11	5	0
480	490	63	56	50	44	38	32	25	19	13	7	0
490	500	64	58	52	45	39	33	27	21	14	8	2
500	510	66	59	53	47	41	35	28	22	16	10	3
510	520	67	61	55	48	42	36	30	24	17	11	5
520	530	69	62	56	50	44	38	31	25	19	13	6
530	540	70	64	58	51	45	39	33	27	20	14	8
540	550	72	65	59	53	47	41	34	28	22	16	9
550	560	73	67	61	54	48	42	36	30	23	17	11
560	570	75	68	62	56	50	44	37	31	25	19	12
570	580	76	70	64	57	51	45	39	33	26	20	14
580	590	78	71	65	59	53	47	40	34	28	22	15
590	600	79	73	67	60	54	48	42	36	29	23	17
600	610	81	74	68	62	56	50	43	37	31	25	18
610	620	82	76	70	63	57	51	45	39	32	26	20
620	630	84	77	71	65	59	53	46	40	34	28	21

USING THE 3 CS

Use the tables on pages 21 and 22 to figure the federal withholding tax for Peggy Laskowitz. Record the tax in the payroll register.

COMPREHEND

In the space below, write what you are to do.

Figure the FWT for Peggy Laskowitz and record it in the payroll register.

Write the steps you should follow to solve the problem.

1. Find the FWT in the tax tables.
2. Record the tax.

Write the information you need to solve the problem.

Gross pay is $340; she is married ("MAR STAT" column says "M"); she claims 4 allowances ("ALLOW" column shows "4").

Write the system of measurement.

Money

COMPUTE

Do the computation required to complete each task.
Use the method above.

In the Married Persons Weekly Payroll Period table, the correct row is for $330-$340. (If her pay were $340.01 or more, you would use the row for $340-$350.) The payroll deduction for 4 allowances is $15.

COMMUNICATE

Enter the information in the payroll register below.

PAYROLL REGISTER   Week Ending 1/10/9X

| EMPLOYEE INFORMATION | | | GROSS EARNINGS | | | DEDUCTIONS | | |
NAME	MAR STAT	ALLOW	TOTAL HOURS	REG RATE	GROSS PAY	FICA	FWT	NET PAY
Peggy Laskowitz	M	4	40	8.50	340.00	26.01	15.00	
Angela Upton	S	1	40	7.75	310.00	23.72		
Patrick O'Leary	S	1	39	10.70	417.30	31.92		
Jamal Cheston	M	5	25	11.00	275.00	21.04		
Barbara Palmer	S	2	43	8.25	354.75	27.14		
Sam Li	M	3	39 ½	9.20	363.40	27.80		

PRACTICE

11 Find the federal withholding tax for Angela Upton. Record the tax in the payroll register.

COMPREHEND

In the space below, write what you are to do.

_____

_____

Write the steps you should follow to solve the problem.

_____

_____

Write the information you need to solve the problem.

_____

Write the system of measurement.

_____

COMPUTE

Do the computation required to complete each task.
Use the method given on page 23.

_____

COMMUNICATE

Enter the information in the payroll register on page 23.

12 Find the federal withholding tax and take-home pay for Patrick O'Leary. Record the tax in the payroll register.

COMPREHEND

In the space below, write what you are to do.

_____

_____

Write the steps you should follow to solve the problem.

_____

_____

Write the information you need to solve the problem.

_____

Write the system of measurement.

_____

COMPUTE
Do the computation required to complete each task.
Use the method given on page 23.

_____

COMMUNICATE
Enter the information in the payroll register on page 23.

13 Find the federal withholding tax and take-home pay for Jamal Cheston.
Record the tax in the payroll register.

COMPREHEND
In the space below, write what you are to do.

_____

_____

Write the steps you should follow to solve the problem.

_____

_____

Write the information you need to solve the problem.

_____

Write the system of measurement.

_____

COMPUTE
Do the computation required to complete each task.
Use the method given on page 23.

_____

COMMUNICATE
Enter the information in the payroll register on page 23.

14 Find the federal withholding tax and take-home pay for Barbara Palmer.
Record the tax in the payroll register.

COMPREHEND
In the space below, write what you are to do.

_____

_____

Write the steps you should follow to solve the problem.

_____

_____

Write the information you need to solve the problem.

_____

Write the system of measurement.

_____

COMPUTE
Do the computation required to complete each task.
Use the method given on page 23.

_____

COMMUNICATE
Enter the information in the payroll register on page 23.

15 Find the federal withholding tax and take-home pay for Sam Li. Record the tax in the payroll register.

COMPREHEND
In the space below, write what you are to do.

_____

_____

Write the steps you should follow to solve the problem.

_____

_____

Write the information you need to solve the problem.

_____

Write the system of measurement.

_____

COMPUTE
Do the computation required to complete each task.
Use the method given on page 23.

_____

COMMUNICATE
Enter the information in the payroll register on page 23.

TASK 4 FIGURING TAKE-HOME PAY

Once you have figured all the deductions, you are ready to find each employees take-home (or net) pay. This is the amount you will put on the paycheck.

To figure an employee's take-home pay, follow these steps:

1. Figure the total deductions. Use this formula:

FICA + FWT = total deductions

2. Subtract the total deductions from the gross pay to get the net or take-home pay.

gross pay − total deductions = take-home pay

USING THE 3 CS

Use the payroll register below to figure the take-home pay for Peggy Laskowitz. Record the net pay in the payroll register.

COMPREHEND

In the space below, write what you are to do.

Figure the take-home pay for Peggy Laskowitz and record it in the payroll register.

Write the steps you should follow to solve the problem.

1. Add the FICA and FWT shown in the payroll register to get the total deductions.
2. Subtract total deductions from the gross pay.

Write the information you need to solve the problem.

Gross pay is $340.00. FICA is $26.01. FWT is $15.00.

Write the system of measurement.

Money

COMPUTE

Do the computation required to complete each task.
Use the formulas given above.

FICA + FWT = total deductions
$26.01 + $15.00 = $41.01

gross pay − total deductions = net pay
$340 − $41.01 = $298.99

COMMUNICATE

Enter the information in the payroll register below.

PAYROLL REGISTER   Week Ending 1/10/9X

NAME	MAR STAT	ALLOW	TOTAL HOURS	REG RATE	GROSS PAY	FICA	FWT	NET PAY
	EMPLOYEE INFORMATION			GROSS EARNINGS		DEDUCTIONS		
Peggy Laskowitz	M	4	40	8.50	340.00	26.01	15.00	298.99
Angela Upton	S	1	40	7.75	310.00	23.72	36.00	
Patrick O'Leary	S	1	39	10.70	417.30	31.92	52.00	
Jamal Cheston	M	5	25	11.00	275.00	21.04	0.00	
Barbara Palmer	S	2	43	8.25	354.75	27.14	37.00	
Sam Li	M	3	39½	9.20	363.40	27.80	26.00	

PRACTICE

16 Find the take-home pay for Angela Upton. Record the take-home pay in the payroll register.

COMPREHEND
In the space below, write what you are to do.

_____

_____

Write the steps you should follow to solve the problem.

_____

_____

Write the information you need to solve the problem.

_____

_____

Write the system of measurement.

_____

COMPUTE
Do the computation required to complete each task.
Use the formulas given on page 27.

_____

_____

COMMUNICATE
Enter the information in the payroll register above.

17 Find the take-home pay for Patrick O'Leary. Record the take-home pay in the payroll register.

COMPREHEND
In the space below, write what you are to do.

_____

_____

Write the steps you should follow to solve the problem.

_____

_____

Write the information you need to solve the problem.

_____

_____

Write the system of measurement.

_____

COMPUTE
Do the computation required to complete each task.
Use the formulas given on page 27.

_____

_____

COMMUNICATE
Enter the information in the payroll register on page 28.

18 Find the take-home pay for Jamal Cheston. Record the take-home pay in the payroll register.

COMPREHEND
In the space below, write what you are to do.

_____

_____

Write the steps you should follow to solve the problem.

_____

_____

Write the information you need to solve the problem.

_____

_____

Write the system of measurement.

_____

COMPUTE
Do the computation required to complete each task.
Use the formulas given on page 27.

_____

_____

COMMUNICATE
Enter the information in the payroll register on page 28.

19 Find the take-home pay for Barbara Palmer. Record the take-home pay in the payroll register.

COMPREHEND
In the space below, write what you are to do.

_____

_____

Write the steps you should follow to solve the problem.

_____

_____

Write the information you need to solve the problem.

_____

_____

Write the system of measurement.

_____

COMPUTE
Do the computation required to complete each task.
Use the formulas given on page 27.

_____

_____

COMMUNICATE
Enter the information in the payroll register on page 28.

20 Find the take-home pay for Sam Li. Record the take-home pay in the payroll register.

COMPREHEND
In the space below, write what you are to do.

_____

_____

Write the steps you should follow to solve the problem.

_____

_____

Write the information you need to solve the problem.

_____

_____

Write the system of measurement.

_____

COMPUTE
Do the computation required to complete each task.
Use the formulas given on page 27.

_____

_____

COMMUNICATE
Enter the information in the payroll register on page 28.

❑ Job Situation 2 Estimates and Billing

When customers come to the A&P Auto Center, the first person they talk to is the service manager, Conchita Perez. Conchita finds out what problem the customer is having. She fills out a **work order**. A work order is shown in Figure 2-1 of the *Knowledge Base*. It tells what work should be done on the car. On busy days, you work as Conchita's assistant at the service desk. You fill out work orders using the information you get from Conchita and the customer.

First you write information about the customer and the car. You need the owner's name and address. You also write the car's make, model, and license number.

Conchita tells you the work to be done on the car. You list it. For example, the work order in Figure 2-1 of the *Knowledge Base* shows that the car should get a tune-up. It also shows that the technician should change the oil and filter and lubricate the car. The work order lists parts needed for the job. For example, *oil filter* is written in the "Part Description" section. Figure 3-18 of the *Knowledge Base* shows the prices of various parts.

Then you write an **estimate** of the cost. An estimate is the amount the shop expects to the charge the customer. The cost is made up of two parts:

❑ Cost of labor to do the work. A charge is made for every half hour of work on the car.
❑ Cost of parts needed to do the repair. Each part has its own cost.

Conchita will give you an estimate of the number of hours. You will calculate the cost. Then you look up the part in a price list to get its cost.

Estimates protect both the shop and the customer. The customer knows what he or she will have to pay. The shop knows that the customer agrees to pay. Conchita either has the customer sign the work order or gets phone approval of the estimate.

Sometimes, it's very easy to tell what work needs to be done. Other times, Conchita has to listen carefully to the problems the owner reports. She will write down what the problem is. For example, she may write *Brakes squeal and seem weak* or *Stalls at intersections*. She has to decide what is the most likely cause of the problem. She may check the car quickly herself. Then she may write *Check brakes* or *Check battery* on the work order.

When a technician checks the car and finds the problem, you complete the estimate. You will put down the estimated cost of labor and parts. Then Conchita will call the owner. She will tell the owner what needs to be repaired. She will give the estimate. The owner will approve the work over the phone.

Once the owner has agreed to the estimate, the technician can do the work. When the technician finishes, he or she gives the work order back you. Attached is a list of the parts used. The technician tells how much time he or she spent. Now it's your job to turn the work order into the **final bill**.

❑ You find the price for any new parts.
❑ You find the actual cost of labor when it is different from the estimate.
❑ You find the total cost.

The work order in Figure 2-1 of the *Knowledge Base* is also the final bill. It shows the final costs for Mrs. Preston's car. It shows a charge for labor for the tune-up—$69.95. It also shows the parts used. The technician replaced the oil filter. He also replaced the air filter. This wasn't on the original work order, so the cost of this was added. He also used other parts that are listed on the bill. Figure 3-18 of the *Knowledge Base* has the prices for parts.

When the customer comes to pick up his or her car, the bill will be ready. Customers don't like to have to wait for their cars. They appreciate being able to pay and leave quickly.

TASK 1—FILLING OUT THE WORK ORDER

A work order is a legal agreement between the shop and the customer. It tells what work will be done. It says that the car's owner agrees to have that work done and to pay for it.

You need certain pieces of information in order to begin to fill out a work order:

❏ Information about the customer. You get this by talking to the customer or by looking at his or her driver's license. You need the name, address, and phone number.
❏ Information about the car. You get this from the car registration. You need the make, model, year, and license number of the car.
❏ Information about the work to be done. You get this from the service manager.
❏ Information about parts to be used. This comes from the service manager.

Here are the steps you follow:

1. Fill in the upper left corner of the work order. Include information about the customer and the car.
2. Fill in the "Service/Labor" section of the work order. Write a description of the services.
3. Fill in the "Part" section of the work order. List the parts to be used, including the part number if there is one.

In the next job task, you will find and fill in the cost of each item on the work order.

USING THE 3 CS

Mrs. Grossman brings her car in for service. She gives you the following information:

Name: Mrs. J. Grossman
Address: 15 West Elm St.
       Northtown, IL 50050
Phone: 555-1234

You get this information from the car registration:

89 Toyota Tercel

License 1YD 226

Conchita tells you that the technician will install a new muffler and tail pipe. She gives you the parts numbers: muffler—34G, tail pipe—12-67A8.
Write this information in the proper place on the work order.

COMPREHEND
In the space below, write what you are to do.

Fill in the information on the work order.

Write the steps you should follow to solve the problem.

1. Fill in the customer and car information.
2. Fill in the labor description.
3. List the parts.

Write the information you need to solve the problem.

Name: Mrs. J. Grossman
Address:15 West Elm St.
*   Northtown, IL 50050*
Phone: 555-1234

89 Toyota Tercel
License 1YD 226

Labor: install muffler and tail pipe
Parts: muffler—34G; tail pipe—12-67A8.

Write the system of measurement.

Time and money

COMPUTE
Complete each task. Follow the steps listed on page 33.
Put the information where it belongs on the work order below.

COMMUNICATE
Write the information on the work order below.

WORK ORDER

DATE: 1/10/9X				SERVICE/LABOR		PRICE
NAME: Mrs. J. Grossman **ADDRESS:** 15 West Elm St. **CITY/ZIP:** Northtown, IL 50050				Install muffler/tail pipe		
PHONE: 555-1234						
YEAR	MAKE	MODEL	LICENSE			
89	Toyota	Tercel	1YD 226	**TOTAL LABOR**		
QTY.	PART		PRICE	TOTAL LABOR		
1 1	Muffler 34G Tail pipe 12-67A8			TOTAL PARTS TOTAL SALE		
	TOTAL PARTS			**CUSTOMER SIGNATURE:** _____		

PRACTICE

1 Mr. Billings brings his car in for service. He gives you the following information:

Name: Alan Billings
Address: 100 Plymouth Ave.
       Northtown, IL 50050
Phone: 555-3498

You get this information from the car registration:

90 Ford Taurus
License 4RD 111

Conchita tells you that the technician will do a tune-up. She gives you the part number: 6 spark plugs 135A.

Write this information in the proper place on the work order.

COMPREHEND

In the space below, write what you are to do.

_____

Write the steps you should follow to solve the problem.

_____

_____

_____

Write the information you need to solve the problem.

_____

_____

_____

_____

_____

_____

_____

_____

Write the system of measurement.

_____

COMPUTE

Complete each task. Follow the steps listed on page 33.

COMMUNICATE

Write the information on the work order below.

WORK ORDER

DATE:				SERVICE/LABOR		PRICE
NAME: ADDRESS: CITY/ZIP: PHONE:						
YEAR	MAKE	MODEL	LICENSE			
					TOTAL LABOR	
QTY.	PART		PRICE	TOTAL LABOR TOTAL PARTS TOTAL SALE		
	TOTAL PARTS			CUSTOMER SIGNATURE: _____		

2 Ms. Anderson brings her car in for service. She gives you the following information:

>   Name: Alice Anderson
>   Address: 1 Crescent Blvd.
>   Northtown, IL 50050
>   Phone: 555-1111

You get this information from the car registration:

>   82 Chevrolet Caprice
>   License 5TY 123

Conchita tells you that the technician will change the oil. She gives you the parts: oil filter 12J; 4 quarts of oil (no part number).
   Write this information in the proper place on the work order.

COMPREHEND
In the space below, write what you are to do.

_____

Write the steps you should follow to solve the problem.

_____

_____

_____

Write the information you need to solve the problem.

_____

_____

_____

_____

_____

_____

_____

Write the system of measurement.

_____

COMPUTE
Complete each task. Follow the steps listed on page 33.

COMMUNICATE
Write the information on the work order on page 37.

WORK ORDER

DATE: NAME: ADDRESS: CITY/ZIP: PHONE:				SERVICE/LABOR	PRICE
YEAR	**MAKE**	**MODEL**	**LICENSE**		
				TOTAL LABOR	
QTY.	**PART**		**PRICE**	TOTAL LABOR TOTAL PARTS TOTAL SALE	
				CUSTOMER SIGNATURE: _____	
	TOTAL PARTS				

3   Ms. Kirkpatrick brings her car in for service. She gives you the following information:

    Name: Kathryn Kirkpatrick
    Address: 34 Old Mill Lane
              Northtown, IL 50050
    Phone: 555-3498

You get this information from the car registration:

    87 Honda Civic
    License 4GT 222

Conchita tells you that the technician will check the battery. After the technician checks it, she tells you that the technician will replace it with battery UB-111.
    Write this information in the proper place on the work order.

COMPREHEND
In the space below, write what you are to do.

_____

Write the steps you should follow to solve the problem.

_____

_____

_____

Write the information you need to solve the problem.

_____

_____

_____

_____

_____

_____

_____

_____

Write the system of measurement.

_____

COMPUTE
Complete each task. Follow the steps listed on page 33.

COMMUNICATE
Write the information on the work order below.

WORK ORDER

DATE:				SERVICE/LABOR		PRICE
NAME: ADDRESS: CITY/ZIP: PHONE:						
YEAR	MAKE	MODEL	LICENSE			
					TOTAL LABOR	
QTY.	PART			PRICE	TOTAL LABOR TOTAL PARTS TOTAL SALE	
					CUSTOMER SIGNATURE: _____	
	TOTAL PARTS					

4   Mr. Jones brings his car in for service. He gives you the following information:

    Name: Herb Jones
    Address: 111 West Birch
            Northtown, IL 50050

Phone: 555-7634

You get this information from the car registration:

79 Dodge Dart
License 367 UUY

Conchita tells you that the technician will check the water pump. After the technician checks it, she tells you that the technician will replace the pump with part 328Q.
Write this information in the proper place on the work order.

COMPREHEND
In the space below, write what you are to do.

_____

Write the steps you should follow to solve the problem.

_____

_____

_____

Write the information you need to solve the problem.

_____

_____

_____

_____

_____

_____

_____

Write the system of measurement.

_____

COMPUTE
Complete each task. Follow the steps listed on page 33.

COMMUNICATE
Write the information on the work order on page 40.

WORK ORDER

DATE:				SERVICE/LABOR		PRICE
NAME: ADDRESS: CITY/ZIP: PHONE:						
YEAR	MAKE	MODEL	LICENSE			
				TOTAL LABOR		
QTY.	PART		PRICE	TOTAL LABOR TOTAL PARTS TOTAL SALE		
				CUSTOMER SIGNATURE: _____		
	TOTAL PARTS					

5   Mr. Avino brings his car in for service. He gives you the following information:

    Name: Sam Avino
    Address: 527 East Elm
              Northtown, IL 50050
    Phone: 555-1287

You get this information from the car registration:

    88 Buick Regal
    License SAMMY

Conchita tells you that the technician will adjust the brakes.
    Write this information in the proper place on the work order.

COMPREHEND
In the space below, write what you are to do.

_____

_____

Write the steps you should follow to solve the problem.

_____

_____

_____

Write the information you need to solve the problem.

_____

_____

_____

_____

_____

_____

Write the system of measurement.

_____

COMPUTE
Complete each task. Follow the steps listed on page 33.

COMMUNICATE
Write the information on the work order below.

WORK ORDER

DATE:				SERVICE/LABOR		PRICE
NAME:						
ADDRESS:						
CITY/ZIP:						
PHONE:						
YEAR	MAKE	MODEL	LICENSE			
					TOTAL LABOR	

QTY.	PART		PRICE	TOTAL LABOR	
				TOTAL PARTS	
				TOTAL SALE	
	TOTAL PARTS			CUSTOMER SIGNATURE: _____	

TASK 2—COMPLETING THE ESTIMATE

Once you have listed the labor and parts on each work order, you complete the estimate. You write the cost for each item. Conchita lets the customer know how much labor and parts will cost. The customer then agrees to have the work done.

To complete the estimate, follow these steps:

1. Find the labor cost, using this formula:

hours times $35 (hourly rate) = labor cost

2. Record the labor cost.
3. Look up the price of the parts in a price list.

   Use this formula to find the cost of each kind of part:

number of parts times cost per part = cost of parts

4. Record the parts cost.

USING THE 3 CS

Conchita tells you that work on Mrs. Grossman's car will take 1 hour. Use Figure 3-18 in the *Knowledge Base* to get the parts costs.
Write the labor and parts costs on the work order.

COMPREHEND

In the space below, write what you are to do.

> *Fill in the costs on the work order.*

Write the steps you should follow to solve the problem.

> *1. Find and record the labor cost.*
> *2. Find the parts cost in Figure 3-18 of the* Knowledge Base.
> *3. Record the parts cost.*

Write the information you need to solve the problem.

> *Labor: 1 hour*
> *Parts: muffler—34G; tail pipe—12-67A8.*

Write the system of measurement.

> *Time and money*

COMPUTE

Do the computation required to complete each task. Use the formulas above.

> *hour × $35 = labor cost*
> *1 × $35 = $35*
> *muffler—34G: 1 × $58.00 = $58.00*
> *tail pipe—12-67A8: 1 × $13.50 = $13.50*

COMMUNICATE

Write the information on the work order on page 43.

WORK ORDER

DATE: 1/10/9X				SERVICE/LABOR	PRICE
NAME: Mrs. J. Grossman				Install muffler/tail pipe	$35.00
ADDRESS: 15 West Elm St.					
CITY/ZIP: Northtown, IL 50050					
PHONE: 555-1234					

YEAR	MAKE	MODEL	LICENSE	TOTAL LABOR	
89	Toyota	Tercel	1YD 226		

QTY.	PART		PRICE	TOTAL LABOR	
1	Muffler 34G		$58.00	TOTAL PARTS	
1	Tail pipe 12-67A8		$13.50	TOTAL SALE	
	TOTAL PARTS			CUSTOMER SIGNATURE: _____	

PRACTICE

6 Conchita tells you that work on Mr. Billing's car will take one hour. Use the price list shown in Figure 3-18 of the *Knowledge Base*. Write the labor and parts costs on the work order.

COMPREHEND
In the space below, write what you are to do.

_____

Write the steps you should follow to solve the problem.

_____

_____

Write the information you need to solve the problem.

_____

Write the system of measurement.

_____

COMPUTE
Do the computation required to complete each task. Use the formulas on page 42.

COMMUNICATE
Write the information on the work order below.

WORK ORDER

DATE: 1/10/9X				SERVICE/LABOR	PRICE
NAME: Alan Billings				Tune-up	
ADDRESS: 100 Plymouth Ave.					
CITY/ZIP: Northtown, IL 50050					
PHONE: 555-3498					
YEAR	MAKE	MODEL	LICENSE		
90	Ford	Taurus	4RD 111	TOTAL LABOR	
QTY.	PART		PRICE	TOTAL LABOR	
6	Spark plugs 135A			TOTAL PARTS	
				TOTAL SALE	
				CUSTOMER SIGNATURE: _____	
	TOTAL PARTS				

7 | Conchita tells you that work on Ms. Anderson's car will take half an hour. Use the price list shown in Figure 3-18 of the *Knowledge Base*. Write the labor and parts costs on the work order.

COMPREHEND
In the space below, write what you are to do.

_____

Write the steps you should follow to solve the problem.

_____

_____

_____

Write the information you need to solve the problem.

_____

_____

Write the system of measurement.

_____

COMPUTE
Do the computation required to complete each task. Use the formulas on page 42.

COMMUNICATE
Write the information on the work order below.

WORK ORDER

DATE: 1/10/9X				SERVICE/LABOR	PRICE
NAME: Alice Anderson ADDRESS: 1 Crescent Blvd. CITY/ZIP: Northtown, IL 50050				Oil change	
PHONE: 555-1111					
YEAR	MAKE	MODEL	LICENSE		
82	Chevy	Caprice	5TY 123	TOTAL LABOR	

QTY.	PART		PRICE	TOTAL LABOR	
1	Oil filter 12J			TOTAL PARTS	
4	Qt. oil			TOTAL SALE	
	TOTAL PARTS			CUSTOMER SIGNATURE: _____	

8 Conchita tells you that replacing the battery in Ms. Kirkpatricks's car will take half an hour. There is no charge for checking the battery. (Write *n/c* on the work order.) Use the price list shown in Figure 3-18 of the *Knowledge Base*. Write the labor and parts costs on the work order.

COMPREHEND
In the space below, write what you are to do.

_____

Write the steps you should follow to solve the problem.

_____

_____

_____

Write the information you need to solve the problem.

_____

_____

Write the system of measurement.

_____

COMPUTE
Do the computation required to complete each task. Use the formulas on page 42.

COMMUNICATE
Write the information on the work order below.

WORK ORDER

DATE: 1/10/9X				SERVICE/LABOR	PRICE
NAME: Kathryn Kirkpatrick **ADDRESS:** 34 Old Mill Lane **CITY/ZIP:** Northtown, IL 50050 **PHONE:** 555-3498				Check battery Replace battery	
YEAR	**MAKE**	**MODEL**	**LICENSE**		
87	Honda	Civic	4GT 222	**TOTAL LABOR**	
QTY.	**PART**		**PRICE**	**TOTAL LABOR**	
1	Battery UB-111			**TOTAL PARTS**	
				TOTAL SALE	
	TOTAL PARTS			**CUSTOMER SIGNATURE:** _____	

[9] Conchita tells you that work on Mr. Jones's car will take 2½ hours. Use the price list shown in Figure 3-18 of the *Knowledge Base*. Write the labor and parts costs on the work order.

COMPREHEND
In the space below, write what you are to do.

_____

Write the steps you should follow to solve the problem.

_____

_____

_____

Write the information you need to solve the problem.

_____

_____

Write the system of measurement.

_____

COMPUTE
Do the computation required to complete each task. Use the formulas on page 42.

COMMUNICATE
Write the information on the work order below.

WORK ORDER

DATE: 1/10/9X				SERVICE/LABOR	PRICE
NAME: Herb Jones ADDRESS: 111 West Birch CITY/ZIP: Northtown, IL 50050 PHONE: 555-7634				Replace water pump	

YEAR	MAKE	MODEL	LICENSE	TOTAL LABOR	
79	Dodge	Dart	367 UUY		

QTY.	PART		PRICE	TOTAL LABOR TOTAL PARTS TOTAL SALE	
1	Water pump 328Q				
	TOTAL PARTS			CUSTOMER SIGNATURE: _____	

10 Conchita tells you that work on Mr. Avino's car will take one hour. Use the price list shown in Figure 3-18 of the *Knowledge Base*. Write the labor and parts costs on the work order.

COMPREHEND
In the space below, write what you are to do.

_____

Write the steps you should follow to solve the problem.

_____

_____

Write the information you need to solve the problem.

_____

_____

Write the system of measurement.

_____

COMPUTE
Do the computation required to complete each task. Use the formulas on page 42.

COMMUNICATE
Write the information on the work order below.

WORK ORDER

DATE: 1/10/9X				SERVICE/LABOR	PRICE
NAME: Sam Avino				Adjust brakes	
ADDRESS: 527 East Elm					
CITY/ZIP: Northtown, IL 50050					
PHONE: 555-1287					
YEAR	MAKE	MODEL	LICENSE	**TOTAL LABOR**	
88	Buick	Regal	SAMMY		

QTY.	PART		PRICE	TOTAL LABOR	
				TOTAL PARTS	
				TOTAL SALE	
	TOTAL PARTS			**CUSTOMER SIGNATURE:** _____	

TASK 3—COMPLETING THE FINAL BILL

After the work on a car is done, the technician gives you back the work order. If the technician has used any parts not shown on the work order, he or she lists them. If labor took more or less time than shown on the work order, he or she tells you this too.

Now you need to turn the work order into the final bill. Follow these steps:

1. Check the labor cost. Correct it if more or fewer hours of labor were needed.
2. Write the total labor cost next to the heading "Total Labor."
3. Check the parts cost. Write the cost of any new parts on the work order.
4. Add the cost of all the parts. Write the total next to the heading "Total Parts."
5. Copy the total labor and total parts costs to the bottom right section of the work order. Find the total sale, using this formula.

$$\text{total labor} + \text{total parts} = \text{total sale}$$

USING THE 3 Cs
The technician has completed work on Mrs. Grossman's car. Labor costs were the amount that was estimated. The technician used the parts listed on the work order. She also replaced a gasket with part 542. Complete the final bill.

COMPREHEND
In the space below, write what you are to do.

Complete the final bill.

Write the steps you should follow to solve the problem.

> 1. Check the labor cost.
> 2. Write the total labor.
> 3. Check the parts cost. Write the name and cost of any new parts on the work order.
> 4. Add the cost of all the parts.
> 5. Find the total sale.

Write the information you need to solve the problem.

> *Labor costs were as estimated. Gasket 542 was used.*

Write the system of measurement.

> *Time and money.*

COMPUTE
Do the computation required to complete each task. Follow the steps listed on page 48.

> *Labor costs: $35.00 (as shown on work order)*
> *New part: 1 × $2.98 = $2.98*
> *Total parts: $58.00 + $13.50 + $2.98 = $74.48*
> *Total sale: $35.00 + $74.48 = $109.48*

COMMUNICATE
Write the information on the work order below.

WORK ORDER

DATE: 1/10/9X				SERVICE/LABOR	PRICE
NAME: Mrs. J. Grossman **ADDRESS:** 15 West Elm St. **CITY/ZIP:** Northtown, IL 50050				Install muffler/tail pipe	$35.00
PHONE: 555-1234					
YEAR	MAKE	MODEL	LICENSE		
89	Toyota	Tercel	1YD 226	**TOTAL LABOR**	$35.00

QTY.	PART	PRICE		
			TOTAL LABOR	$35.00
1	Muffler 34G	$58.00	TOTAL PARTS	$74.48
1	Tail pipe 12-67A8	$13.50		
1	*Gasket 542*	$ 2.98	TOTAL SALE	$109.48
	TOTAL PARTS	$74.48	**CUSTOMER SIGNATURE:** _____	

11 The technician has completed work on Mr. Billings's car. Labor costs were the amount that was estimated. The technician used the parts listed on the work order. Complete the final bill.

COMPREHEND
In the space below, write what you are to do.

_____

Write the steps you should follow to solve the problem.

_____
_____
_____
_____
_____
_____

Write the information you need to solve the problem.

_____

Write the system of measurement.

_____

COMPUTE
Do the computation required to complete each task. Use the formulas on page 48.

COMMUNICATE
Write the information on the work order below.

WORK ORDER

DATE: 1/10/9X				SERVICE/LABOR	PRICE
NAME: Alan Billings ADDRESS: 100 Plymouth Ave. CITY/ZIP: Northtown, IL 50050				Tune-up	$35.00
PHONE: 555-3498					
YEAR	MAKE	MODEL	LICENSE		
90	Ford	Taurus	4RD 111	TOTAL LABOR	

QTY.	PART		PRICE	TOTAL LABOR	
6	Spark plugs 135A		$21.00	TOTAL PARTS	
				TOTAL SALE	
				CUSTOMER SIGNATURE: _____	
	TOTAL PARTS				

12 The technician has completed work on Ms. Anderson's car. Labor costs were the amount that was estimated. The technician used the parts listed on the work order. He also replaced an air filter with part A26. Complete the final bill.

COMPREHEND
In the space below, write what you are to do.

_____

Write the steps you should follow to solve the problem.

_____

_____

_____

_____

_____

_____

Write the information you need to solve the problem.

_____

_____

Write the system of measurement.

_____

COMPUTE
Do the computation required to complete each task. Use the formulas on page 48.

COMMUNICATE
Write the information on the work order below.

WORK ORDER

DATE: 1/10/9X				SERVICE/LABOR	PRICE
NAME: Alice Anderson ADDRESS: 1 Crescent Blvd. CITY/ZIP: Northtown, IL 50050 PHONE: 555-1111				Oil change	$17.50
YEAR	MAKE	MODEL	LICENSE	TOTAL LABOR	
82	Chevy	Caprice	5TY 123		

QTY.	PART		PRICE	TOTAL LABOR	
1	Oil filter 12J		$6.15	TOTAL PARTS	
4	Qt. oil		$7.04	TOTAL SALE	
	TOTAL PARTS			CUSTOMER SIGNATURE: _____	

13 The technician has completed work on Ms. Kirkpatrick's car. Labor costs were the amount that was estimated. The technician used the parts listed on the work order. He also had to replace 4 spark plugs with part 172F. Complete the final bill.

COMPREHEND
In the space below, write what you are to do.

_____

Write the steps you should follow to solve the problem.

_____
_____
_____
_____
_____
_____

Write the information you need to solve the problem.

_____
_____

Write the system of measurement.

_____

COMPUTE
Do the computation required to complete each task. Use the formulas on page 48.

COMMUNICATE
Write the information on the work order below.

WORK ORDER

DATE: 1/10/9X				SERVICE/LABOR	PRICE
NAME: Kathryn Kirkpatrick				Check battery	n/c
ADDRESS: 34 Old Mill Lane				Replace battery	$17.50
CITY/ZIP: Northtown, IL 50050					
PHONE: 555-3498					
YEAR	MAKE	MODEL	LICENSE	TOTAL LABOR	
87	Honda	Civic	4GT 222		
QTY.	PART		PRICE	TOTAL LABOR	
1	Battery UB-111		$52.00	TOTAL PARTS	
				TOTAL SALE	
				CUSTOMER SIGNATURE: _____	
	TOTAL PARTS				

14 The technician has completed work on Mr. Jones's car. Labor costs were the amount that was estimated. The technician used the parts listed on the work order. She also had to replace a belt with part 472F. Complete the final bill.

COMPREHEND
In the space below, write what you are to do.

_____

Write the steps you should follow to solve the problem.

_____

_____

_____

_____

_____

_____

Write the information you need to solve the problem.

_____

_____

Write the system of measurement.

_____

COMPUTE
Do the computation required to complete each task. Use the formulas on page 48.

COMMUNICATE
Write the information on the work order on page 54.

WORK ORDER

DATE: 1/10/9X	SERVICE/LABOR	PRICE
NAME: Herb Jones ADDRESS: 111 West Birch CITY/ZIP: Northtown, IL 50050 PHONE: 555-7634	Replace water pump	$87.50

YEAR	MAKE	MODEL	LICENSE		
79	Dodge	Dart	367 UUY	TOTAL LABOR	

QTY.	PART	PRICE		
1	Water pump 328Q	$62.27	TOTAL LABOR TOTAL PARTS TOTAL SALE	
			CUSTOMER SIGNATURE: _____	
	TOTAL PARTS			

15 The technician has completed work on Mr. Avino's car. Labor was 1½ hours instead of the amount that was estimated. The technician used the 2 brake pads, part number 673. Complete the final bill.

COMPREHEND
In the space below, write what you are to do.

_____

Write the steps you should follow to solve the problem.

_____

_____

_____

_____

_____

Write the information you need to solve the problem.

_____

Write the system of measurement.

_____

COMPUTE

Do the computation required to complete each task. Use the formulas on page 48.

COMMUNICATE

Write the information on the work order below.

WORK ORDER

DATE: 1/10/9X				SERVICE/LABOR		PRICE
NAME: Sam Avino				Adjust brakes		$35.00
ADDRESS: 527 East Elm						
CITY/ZIP: Northtown, IL 50050						
PHONE: 555-1287						
YEAR	MAKE	MODEL	LICENSE			
88	Buick	Regal	SAMMY	TOTAL LABOR		
QTY.	PART		PRICE	TOTAL LABOR		
				TOTAL PARTS		
				TOTAL SALE		
	TOTAL PARTS			CUSTOMER SIGNATURE: _____		

PART 5
MATHEMATICS AND MATERIAL RESOURCES

Part 5 covers some material resources you may deal with on the job. There are three job situations in this part of the book: scheduled maintenance, ordering metric parts, and reading a micrometer. Each job situation requires you to do one or more math tasks. Use the 3 Cs to complete each math task.

❏ JOB SITUATION 1   SCHEDULED MAINTENANCE

You work as an assistant to Conchita Perez. Part of your job is to fill out a work order. A work order is shown in Figure 2-1 of the *Knowledge Base*. Part 4, Job Situation 2—Estimates and Billing tells about work orders.

Some cars have a problem that needs to be **repaired** or fixed. Perhaps the brakes are not working properly. Or the engine is making a noise.

Other cars need **routine maintenance**. Routine maintenance is work done to keep a car running smoothly. It includes changing the oil, replacing filters, and replacing parts that wear out. Car manufacturers set a **maintenance schedule** for each car. They tell the car owner when to bring the car in for maintenance. The schedule tells what services to have done. This information is in the car **owner's manual**. An example of a maintenance schedule is shown in Figure 3-17 of the *Knowledge Base*.

Car manufacturers schedule maintenance at regular periods. The periods are miles. So they may schedule maintenance every 5000 miles or every 7500 miles. If the car should be serviced every 7500 miles, the owner will bring it in at these times: 7500 miles, 15,000 miles, 22,500 miles, 30,000 miles, and so on.

A car owner brings in a car for scheduled maintenance. It's your job to look up the maintenance schedule for that car. You have a complete set of manuals for all the cars you service. You look up the car's mileage and read the chart to tell what service is needed. You record the information on the work order. Then you go on to figure the cost. Figuring costs is explained in Job Situation 2 of Part 4.

TASK 1—USING THE OWNER'S MANUAL

Information on scheduled maintenance is in the car owner's manual. It is given in charts like the one shown in Figure 3-17 of the *Knowledge Base*. Figure 3-17 is for mileage from 0 to 120,000 miles.

Notice that across the top of the chart there are numbers showing mileage. The mileage is shown in thousands of miles.

- ❏ 7.5 means 7.5 thousand miles or 7500 miles.
- ❏ 15 means 15 thousand miles or 15,000 miles.

In the rows in each chart are types of service. For example, the first row shows "change engine oil."

To use the chart, follow these steps:

1. Find the car mileage. If the car is a little over or under a mileage, use that mileage. For example, if a car owner brings a car in at 14,500 miles, look up 15,000 miles.
2. Read down the column to see what service is needed at that mileage. If the service is needed, there will be an *X* in the box.

USING THE 3 CS

Mr. Rivera brings in his car for routine maintenance. The car mileage is 29,900 miles. Find out what maintenance his car needs. Write it on the work order. Use the Maintenance Schedule in Figure 3-17 of the *Knowledge Base*.

COMPREHEND

In the space below, write what you are to do.

Find out what maintenance the car needs.

Write the steps you should follow to solve the problem.

1. Find the car's mileage.
2. Read down the column to find the services needed.

Write the information you need to solve the problem.

Car mileage: 29,900

Write the system of measurement.

Miles

COMPUTE

Do the computation required to complete each task. Follow the steps listed above.

Use the "30,000 miles" column (labeled "30"). There are Xs next to "change engine oil," "replace oil filter," "inspect/adjust drive belts," "replace spark plugs," "replace air filter," "apply solvent to choke shaft," and "check choke heat source."

COMMUNICATE
Write the information in the "Service" section of the work order.

SERVICE/LABOR	PRICE
change oil	
replace oil filter	
inspect/adjust drive belts	
replace spark plugs	
replace air filter	
apply solvent to choke shaft	
check choke heat source	

PRACTICE

1 Mr. Johnson brings in his car for routine maintenance. The car mileage is 7500 miles. Find out what maintenance his car needs. Write it on the work order. Use the Maintenance Schedule in Figure 3-17 of the *Knowledge Base*.

COMPREHEND
In the space below, write what you are to do.

_____

Write the steps you should follow to solve the problem.

_____

Write the information you need to solve the problem.

_____

Write the system of measurement.

_____

COMPUTE
Do the computation required to complete each task. Follow the steps listed on page 58.

_____

COMMUNICATE
Write the information on the work order below.

SERVICE/LABOR	PRICE

2 Mrs. Olenska brings in her car for routine maintenance. The car mileage is 15,100 miles. Find out what maintenance the car needs. Write it on the work order. Use the Maintenance Schedule in Figure 3-17 of the *Knowledge Base*.

COMPREHEND
In the space below, write what you are to do.

_____

Write the steps you should follow to solve the problem.

_____

_____

Write the information you need to solve the problem.

_____

Write the system of measurement.

_____

COMPUTE
Do the computation required to complete each task. Follow the steps listed on page 58.

_____

_____

COMMUNICATE
Write the information on the work order below.

SERVICE/LABOR	PRICE

3 Ms. Wheatley brings in her car for routine maintenance. The car mileage is 89,121 miles. Find out what maintenance the car needs. Write it on the work order. Use the Maintenance Schedule in Figure 3-17 of the *Knowledge Base*.

COMPREHEND
In the space below, write what you are to do.

_____

Write the steps you should follow to solve the problem.

_____

_____

Write the information you need to solve the problem.

_____

Write the system of measurement.

_____

COMPUTE
Do the computation required to complete each task. Follow the steps listed on page 58.

_____

_____

_____

_____

COMMUNICATE
Write the information on the work order below.

SERVICE/LABOR	PRICE

4 Ms. Mona brings in her car for routine maintenance. The car mileage is 59,431 miles. Find out what maintenance the car needs. Write it on the work order. Use the Maintenance Schedule in Figure 3-17 of the *Knowledge Base*.

COMPREHEND
In the space below, write what you are to do.

_____

Write the steps you should follow to solve the problem.

_____

_____

Write the information you need to solve the problem.

_____

Write the system of measurement.

_____

COMPUTE
Do the computation required to complete each task. Follow the steps listed on page 58.

_____

_____

_____

_____

_____

_____

COMMUNICATE
Write the information on the work order below.

SERVICE/LABOR	PRICE

5 Mr. Hill brings in his car for routine maintenance. The car mileage is 53,000 miles. Find out what maintenance the car needs. Write it on the work order. Use the Maintenance Schedule in Figure 3-17 of the *Knowledge Base*.

COMPREHEND
In the space below, write what you are to do.

_____

Write the steps you should follow to solve the problem.

_____

_____

Write the information you need to solve the problem.

_____

Write the system of measurement.

_____

COMPUTE
Do the computation required to complete each task. Follow the steps listed on page 58.

_____

_____

_____

_____

COMMUNICATE
Write the information on the work order below.

SERVICE/LABOR	PRICE

❏ JOB SITUATION 2   ORDERING METRIC PARTS

It's your job to help Patrick order parts for the shop. The shop has to keep many supplies on hand. These include parts like oil filters for different cars. Supplies also include screws, bolts, and other parts. The shop uses these in many repair jobs.

Each month you check supplies to see which items you need. You make a list. Then you check your suppliers **parts lists**. The parts lists tell what parts your suppliers have for sale. They give the price for each item.

To order some parts, you need to know about the **metric system**. The metric system is a system of measurement. It includes units like meters, grams, and liters. Most countries use the metric system. In the United States, we use the **U.S. customary system**. It includes feet, pounds, and quarts. When the technicians in your shop work on foreign cars, they need metric parts. So you need to understand these measurements. Figure 3-16 in the *Knowledge Base* shows how metric and customary measurements compare.

When you look for a metric part in a parts list, you may find that it doesn't show the unit of measurement you are looking for. For example, you may be looking for a 1.5-centimeter washer. The catalog shows washers in millimeters. You have to convert centimeters to millimeters to find the correct part.

Once you decide what parts you need, you fill out a **purchase order**. A purchase order is a form that tells a supplier what parts you want. The purchase orders have A&P Auto's name and address printed on them. They contain spaces for recording information about an order.

You fill in the item names, quantities, and prices. The **unit price** comes from the price list. The unit price is the price for one item. It may be the price for one tire. It may also be the price for *one dozen* screws. Then you figure the **extension price**. The extension price is the price for the number of items you want. You add the prices for all the items you want to buy and write the total cost. Then you mail the purchase order to the supplier.

TASK 1—CONVERTING METRIC MEASUREMENTS

The metric system, like our system of money, is based on tens. Each type of unit is 10, 100, or 1000 times greater or smaller than another unit.

The metric unit of length is the **meter**. A meter is 39.37 inches. It is a little more than a yard. Here are other measures of length.

millimeter

centimeter

kilometer

Notice that each has the word *meter* in it. The word part at the beginning of each word has a specific meaning:

milli = $\frac{1}{1000}$

centi = $\frac{1}{100}$

kilo = 1000

So here are what the words mean:

millimeter (mm) = $\frac{1}{1000}$ meter

centimeter (cm) = $\frac{1}{100}$ meter

kilometer (km) = 1000 meters

You can think of these measurements on a scale from smallest to largest.

Here are some ways that these units compare:

UNIT	EQUAL TO
1 meter (m)	100 centimeters (cm)
1 meter (m)	1000 millimeters (mm)
1 centimeter (cm)	10 millimeters (mm)
1 kilometer (km)	1000 meters (m)

It is easy to change one metric measurement of length to another. Follow these rules:

> To convert a smaller unit to a larger unit, divide. (Think: There are *fewer* meters than millimeters in a length. So you divide to get a *smaller* number.)

> To convert a unit to a smaller unit, multiply. (Think: There are *more* millimeters than meters in a length. So you multiply to get a *larger* number.)

These are the steps you follow:

Example: Convert 5000 millimeters to meters.

1. Find the correct rule:
   To convert a smaller unit to a larger unit, divide.
2. Look in the table above to see how the two units compare.
   m = 1000-mm
3. Do the computation.
   mm ÷ 1000 = m
   5000-mm ÷ 1000 = 5 m

USING THE 3 Cs

You need to order 3 dozen washers that each measure 1.5 centimeters. This is what the parts list shows:

PART	PRICE
15-mm washer	$1.89 doz
150-mm washer	$2.27 doz
30-mm washer	$1.27 doz

Convert a measurement in centimeters to one in millimeters. Decide which part you need to order. Write the quantity, the part name, and the price on the purchase order.

COMPREHEND
In the space below, write what you are to do.

Convert centimeters to millimeters. Find the correct part.

Write the steps you should follow to solve the problem.

1. Find the correct rule.
2. Look in the table to see how the two units compare.
3. Do the computation.

Write the information you need to solve the problem.

Need: 1.5-cm washer

Write the system of measurement.

Length/Metric system

COMPUTE
Do the computation required to complete each task. Follow the steps listed on page 65.

Use the rule: To convert a larger unit to a smaller unit, multiply.
Table shows: 1 cm = 10-mm
  1.5 cm × 10 = 15 mm (Remember, to multiply by 10, move the decimal point one place to the right.)

COMMUNICATE
Write the information on the purchase order below. Notice that you've already written other items on the purchase order.

Supplier: Ed's Auto Parts	Ship to:
Address: 555 West 10th St. Northtown, IL 50050	A&P Auto Center 326 Main St. Chester, IA 40092

Date: 1/15/9X

Quantity	Description of Part	Unit Price	Extension Price
2 doz	oil filter 56JP	$65.00/doz	
1 doz	air filter 88Y	$71.90/doz	
3 doz	15-mm washer	$ 1.89/doz	
		TOTAL:	

PRACTICE

1 You need to order 1 dozen bolts that each measure 1 centimeter. This is what the parts list shows:

PART	PRICE
100-mm bolt	$6.21 doz
50-mm bolt	$3.39 doz
10-mm bolt	$2.90 doz

Convert a measurement in centimeters to one in millimeters. Decide which part you need to order. Write the quantity, the part name, and the price on the purchase order.

COMPREHEND
In the space below, write what you are to do.

_____

Write the steps you should follow to solve the problem.

_____

_____

Write the information you need to solve the problem.

_____

Write the system of measurement.

_____

COMPUTE
Do the computation required to complete each task. Follow the steps listed on page 65.

COMMUNICATE
Write the information on the purchase order below.

Supplier: Ed's Auto Parts		**Ship to:**	
Address: 555 West 10th St. Northtown, IL 50050		A&P Auto Center 326 Main St. Chester, IA 40092	
Date: 1/15/9X			
Quantity	**Description of Part**	**Unit Price**	**Extension Price**
		TOTAL:	

2. You need to order 2 dozen washers that each measure 35 millimeters. This is what the parts list shows:

PART	PRICE
0.35-cm washer	$2.24 doz
3.5-cm washer	$3.26 doz
7-cm washer	$4.18 doz

Convert a measurement in millimeters to one in centimeters. Decide which part you need to order. Write the quantity, the part name, and the price on the purchase order.

COMPREHEND
In the space below, write what you are to do.

_____

Write the steps you should follow to solve the problem.

_____

_____

_____

Write the information you need to solve the problem.

_____

Write the system of measurement.

_____

COMPUTE
Do the computation required to complete each task. Follow the steps listed on page 65.

COMMUNICATE
Write the information on the purchase order below.

Supplier: Ed's Auto Parts **Address:** 555 West 10th St. Northtown, IL 50050		**Ship to:** **A&P Auto Center** 326 Main St. Chester, IA 40092	
Date: 1/15/9X			
Quantity	Description of Part	Unit Price	Extension Price
1 doz	Belt TH-35	$34.00/doz	
		TOTAL:	

3 You need to order 4 dozen bolts that each measure 125 millimeters. This is what the parts list shows:

PART	PRICE
1.25-cm bolt	$1.87 doz
12.5-cm bolt	$3.89 doz
125-cm bolt	$7.56 doz

Convert a measurement in millimeters to one in centimeters. Decide which part you need to order. Write the quantity, the part name, and the price on the purchase order.

COMPREHEND
In the space below, write what you are to do.

_____

Write the steps you should follow to solve the problem.

_____

_____

_____

Write the information you need to solve the problem.

_____

Write the system of measurement.

_____

COMPUTE
Do the computation required to complete each task. Follow the steps listed on page 65.

COMMUNICATE
Write the information on the purchase order below.

Supplier: Ed's Auto Parts		Ship to:	
Address: 555 West 10th St. Northtown, IL 50050		A&P Auto Center 326 Main St. Chester, IA 40092	
Date: 1/15/9X			
Quantity	Description of Part	Unit Price	Extension Price
2 doz	Hose 78-t1	$27.89/doz	
		TOTAL:	

4 | You need to order 2 dozen rods that each measure 1.10 meters. This is what the parts list shows:

PART	PRICE
100-cm rod	$30.25 doz
110-cm rod	$35.22 doz
1100-cm rod	$56.66 doz

Convert a measurement in meters to one in centimeters. Decide which part you need to order. Write the quantity, the part name, and the price on the purchase order.

COMPREHEND
In the space below, write what you are to do.

_____

Write the steps you should follow to solve the problem.

_____

_____

_____

Write the information you need to solve the problem.

_____

Write the system of measurement.

_____

COMPUTE
Do the computation required to complete each task. Follow the steps listed on page 65.

COMMUNICATE
Write the information on the purchase order below.

Supplier: Ed's Auto Parts **Address:** 555 West 10th St. Northtown, IL 50050		**Ship to:** **A&P Auto Center** 326 Main St. Chester, IA 40092	
Date: 1/15/9X			
Quantity	**Description of Part**	**Unit Price**	**Extension Price**
		TOTAL:	

5 You need to order 2 dozen lengths of plastic tubing that each measure 85 centimeters. This is what the parts list shows:

PART	PRICE
0.85-m tubing	$ 5.67 doz
8.5-m tubing	$ 49.75 doz
85-m tubing	$125.99 doz

Convert a measurement in centimeters to one in meters. Decide which part you need to order. Write the quantity, the part name, and the price on the purchase order.

COMPREHEND
In the space below, write what you are to do.

_____

Write the steps you should follow to solve the problem.

_____

_____

_____

Write the information you need to solve the problem.

_____

Write the system of measurement.

_____

COMPUTE
Do the computation required to complete each task. Follow the steps listed on page 65.

COMMUNICATE
Write the information on the purchase order below.

Supplier: Ed's Auto Parts		Ship to: A&P Auto Center	
Address: 555 West 10th St. Northtown, IL 50050		326 Main St. Chester, IA 40092	
Date: 1/15/9X			
Quantity	Description of Part	Unit Price	Extension Price
1 doz	Oil filter 89AB	$67.80/doz	
2 doz	Tire valve 8J	$24.00/doz	
		TOTAL:	

TASK 2—COMPLETING THE PURCHASE ORDER

After you write the item or items you want to order on the purchase order, you have to write the extension price and the total price.

Follow these steps:

1. Find each extension price. Using this formula:

order quantity × unit price = extension price

2. Write each extension price.
3. Find the total price. Add the cost of all the items on the purchase order.

extension price + extension price + ... = Total Price

4. Write the total price.

USING THE 3 CS

Complete the purchase order. Write each extension price and the total price.

COMPREHEND

In the space below, write what you are to do.

Complete the purchase order.

Write the steps you should follow to solve the problem.

1. Find and write each extension price.
2. Find and write the total price.

Write the information you need to solve the problem.

Quantities and prices are shown on the purchase order.

2 doz	oil filter 56JP	$65.00/doz
1 doz	air filter 88Y	$71.90/doz
3 doz	15-mm washer	$ 1.89/doz

Write the system of measurement.

Money

COMPUTE

Do the computation required to complete each task. Follow the steps listed above.

$$2 \times \$65.00 = \$130.00$$
$$1 \times \$71.90 = \$71.90$$
$$3 \times \$1.89 = \$5.67$$
$$\$130.00 + \$71.90 + \$5.67 = \$207.57$$

COMMUNICATE

Write the information on the purchase order below.

Quantity	Description of Part	Unit Price	Extension Price
2 doz	oil filters 56JP	$65.00/doz	$130.00
1 doz	air filters 88Y	$71.90/doz	$ 71.90
3 doz	15 mm washer	$ 1.89/doz	$ 5.67
		TOTAL:	$207.57

PRACTICE

6 Complete the purchase order. Write each extension price and the total price.

COMPREHEND
In the space below, write what you are to do.

_____

Write the steps you should follow to solve the problem.

_____

_____

Write the information you need to solve the problem.

_____

Write the system of measurement.

_____

COMPUTE
Do the computation required to complete each task. Follow the steps listed on page 72.

COMMUNICATE
Write the information on the purchase order below.

Supplier: Ed's Auto Parts		**Ship to:** A&P Auto Center	
Address: 555 West 10th St. Northtown, IL 50050		326 Main St. Chester, IA 40092	
Date: 1/15/9X			
Quantity	**Description of Part**	**Unit Price**	**Extension Price**
1 doz	10-mm bolt	$2.90/doz	
		TOTAL:	

7 Complete the purchase order. Write each extension price and the total price.

COMPREHEND
In the space below, write what you are to do.

_____

Write the steps you should follow to solve the problem.

_____

_____

Write the information you need to solve the problem.

_____

_____

Write the system of measurement.

_____

COMPUTE
Do the computation required to complete each task. Follow the steps listed on page 72.

COMMUNICATE
Write the information on the purchase order below.

Supplier: Ed's Auto Parts		Ship to:	
Address: 555 West 10th St. Northtown, IL 50050		A&P Auto Center 326 Main St. Chester, IA 40092	

Date: 1/15/9X			
Quantity	Description of Part	Unit Price	Extension Price
1 doz	Belt TH-35	$34.00/doz	
2 doz	3.5-cm washer	$ 3.26/doz	
		TOTAL:	

8 Complete the purchase order. Write each extension price and the total price.

COMPREHEND
In the space below, write what you are to do.

_____

Write the steps you should follow to solve the problem.

_____

_____

Write the information you need to solve the problem.

_____

_____

Write the system of measurement.

_____

COMPUTE
Do the computation required to complete each task. Follow the steps listed on page 72.

COMMUNICATE
Write the information on the purchase order below.

Supplier: Ed's Auto Parts		Ship to:	
Address: 555 West 10th St. Northtown, IL 50050		A&P Auto Center 326 Main St. Chester, IA 40092	
Date: 1/15/9X			
Quantity	Description of Part	Unit Price	Extension Price
2 doz	Hose 78-t1	$27.89/doz	
4 doz	12.5-cm bolt	$ 3.89/doz	
		TOTAL:	

9 Complete the purchase order. Write each extension price and the total price.

COMPREHEND
In the space below, write what you are to do.

_____

Write the steps you should follow to solve the problem.

_____

_____

Write the information you need to solve the problem.

_____

Write the system of measurement.

_____

COMPUTE
Do the computation required to complete each task. Follow the steps listed on page 72.

COMMUNICATE
Write the information on the purchase order below.

<table>
<tr><td colspan="3">Supplier: Ed's Auto Parts

Address: 555 West 10th St.
            Northtown, IL 50050</td><td colspan="2">Ship to:
A&P Auto Center
326 Main St.
Chester, IA 40092</td></tr>
<tr><td colspan="5">Date: 1/15/9X</td></tr>
<tr><td>Quantity</td><td>Description of Part</td><td>Unit Price</td><td>Extension Price</td></tr>
<tr><td>2 doz</td><td>110 cm rod</td><td>$35.22/doz</td><td></td></tr>
<tr><td></td><td></td><td></td><td></td></tr>
<tr><td></td><td></td><td>TOTAL:</td><td></td></tr>
</table>

10 | Complete the purchase order. Write each extension price and the total price.

COMPREHEND
In the space below, write what you are to do.

_____

Write the steps you should follow to solve the problem.

_____

_____

Write the information you need to solve the problem.

_____

_____

_____

Write the system of measurement.

_____

COMPUTE
Do the computation required to complete each task. Follow the steps listed on page 72.

COMMUNICATE
Write the information on the purchase order below.

Supplier: Ed's Auto Parts			Ship to: A&P Auto Center 326 Main St. Chester, IA 40092	
Address: 555 West 10th St. Northtown, IL 50050				

Date: 1/15/9X				
Quantity	**Description of Part**		**Unit Price**	**Extension Price**
1 doz	Oil filter 89AB		$67.80/doz	
2 doz	Tire valve 8J		$24.00/doz	
2 doz	0.85 m plastic tubing		$ 5.67/doz	
			TOTAL:	

❏ JOB SITUATION 3 USING A MICROMETER

Your assignment today is to help Sam, the engine repair specialist. When Sam works on an engine, he uses a number of special tools. Some of the tools help him to measure engine parts. Sam uses a number of **micrometers**. A micrometer is a measuring tool that can measure thickness very accurately. It can measure in thousandths or ten-thousandths of an inch.

One kind of micrometer can be used to measure the **diameter** of a rod. The diameter is a line through the center of a circle. This kind of micrometer is called an **outside micrometer**. Figure 5-1 shows an outside micrometer.

To measure with an outside micrometer, tighten the micrometer on the rod. The gauge shows the diameter of the rod.

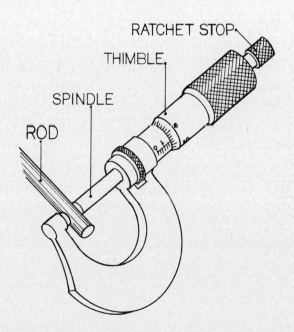

Figure 5-1. An outside micrometer measuring the diameter of a rod.

Micrometers can also be used to measure the size of an opening. They can measure the diameter of an engine cylinder, for example. This kind of micrometer is an **inside micrometer**. Figure 5-2 shows an inside micrometer.

To use an inside micrometer, place it inside the opening to be measured. Rest one end against the side. Then turn the thimble. When the spindle touches the other side, read the diameter.

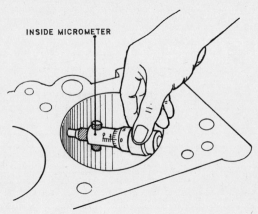

INSIDE MICROMETER

Figure 5-2. An inside micrometer measuring the diameter of a cylinder.

In order to help Sam with his work, you have to be able to read a micrometer. There are two parts to the micrometer gauge.

❏ the thimble
❏ the hub

Figure 5-3 shows the two parts.

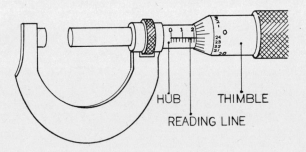

HUB   THIMBLE
READING LINE

Figure 5-3. The micrometer gauge.

The numbers on the hub of this micrometer show tenths of an inch: 0.0 inch, 0.1 inch, 0.2 inch, and so on. In between the numbers are markings that each stand for 25 thousandths of an inch (0.025 inch).

As the thimble is turned, it covers or uncovers markings on the hub. In Figure 5-3, the line for two tenths of an inch (0.2 inch) has been uncovered.

The markings on the thimble stand for smaller lengths. The marks go from 0 to 24. Each number stands for one thousandth of an inch (0.001 inch). When the thimble is turned completely around, it measures 25 thousandths of an inch (0.025 inch).

When the thimble turns completely around, another line on the hub will be uncovered. A complete turn on the thimble stands for 0.025 inch; the line on the hub stands for 0.025 inch.

To read the gauge, add the measurement on the thimble and the measurement on the hub.

TASK 1—READING THE MICROMETER GAUGE

To read a micrometer gauge, like the one shown here, follow these steps:

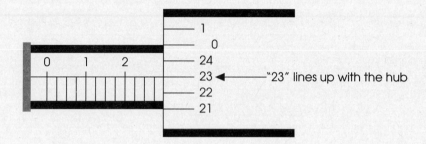

1. Read the number on the hub gauge. Write this measurement. The "2" is uncovered. This stands for 0.2 inch.
2. Read the lines on the hub gauge. Multiply by 0.025. Write this measurement.

   Three marks beyond the "2" are uncovered. Each mark stands for 0.025 inch $3 \times 0.025 = 0.075$ inch. Write this measurement.
3. Read the number on the thimble. Write this measurement.

   The line on the hub points to the "23". This stands for 0.023 inch.
4. Add all the measurements together.

   0.2 in + 0.075 in + 0.023 in =
   $$\begin{array}{r} 0.200 \\ 0.075 \\ 0.023 \\ \hline 0.298 \text{ in} \end{array}$$

USING THE 3 CS

Sam asks you to measure a rod with a micrometer. The gauge is shown here. Read the gauge. Write the measurement.

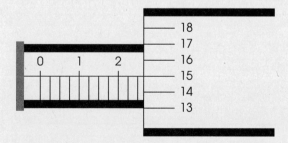

COMPREHEND

In the space below, write what you are to do.

Read the micrometer and write the measurement.

Write the steps you should follow to solve the problem.

> *1. Read the number on the hub.*
> *2. Read the lines on the hub.*
> *3. Read the thimble.*
> *4. Add the measurements.*

Write the information you need to solve the problem.

> *The hub shows "2" and "2" markings. The thimble shows "15."*

Write the system of measurement.

> *Inches*

COMPUTE

Do the computation required to complete each task. Follow the steps listed on page 79.

> | | |
> |---|---|
> | *Hub number:* | *0.2 in* |
> | *Hub lines:* | *2 × .025 = .050 in* |
> | *Thimble:* | *0.015 in* |
>
> *0.2 + 0.050 + 0.015 = 0.265 in*

COMMUNICATE

Write the measurement.

> *0.265 in*

PRACTICE

1 Sam gives you a rod to measure. The gauge is shown here. Read the gauge. Write the measurement.

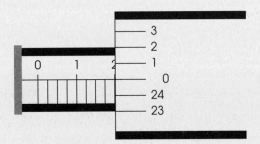

COMPREHEND

In the space below, write what you are to do.

_____

Write the steps you should follow to solve the problem.

_____

_____

_____

_____

Write the information you need to solve the problem.

_____

Write the system of measurement.

_____

COMPUTE
Do the computation required to complete each task. Follow the steps listed on page 79.

COMMUNICATE
Write the measurement.

_____

2. Sam gives you a rod to measure. The gauge is shown here. Read the gauge. Write the measurement.

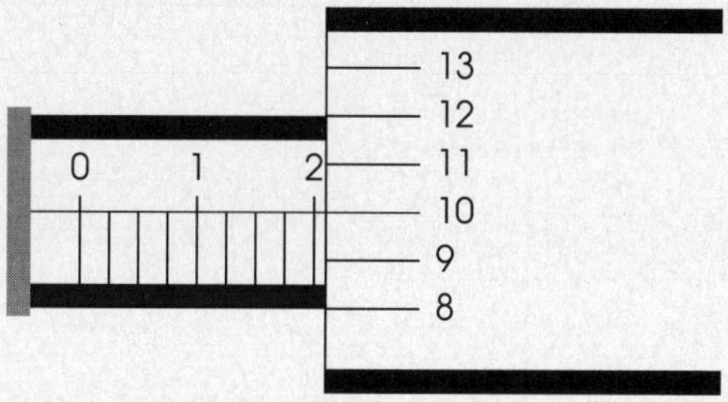

COMPREHEND
In the space below, write what you are to do.

_____

Write the steps you should follow to solve the problem.

_____

_____

_____

_____

Write the information you need to solve the problem.

_____

Write the system of measurement.

_____

COMPUTE

Do the computation required to complete each task. Follow the steps listed on page 79.

COMMUNICATE

Write the measurement.

_____

3 Sam gives you a rod to measure. The gauge is shown here. Read the gauge. Write the measurement.

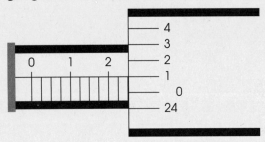

COMPREHEND

In the space below, write what you are to do.

_____

Write the steps you should follow to solve the problem.

_____

_____

_____

_____

Write the information you need to solve the problem.

_____

Write the system of measurement.

_____

COMPUTE

Do the computation required to complete each task. Follow the steps listed on page 79.

COMMUNICATE
Write the measurement.

_____

4 Sam gives you a rod to measure. The gauge is shown here. Read the gauge. Write the measurement.

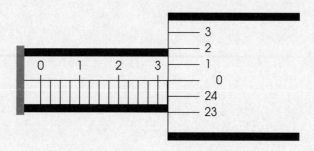

COMPREHEND
In the space below, write what you are to do.

_____

Write the steps you should follow to solve the problem.

_____

_____

_____

Write the information you need to solve the problem.

_____

Write the system of measurement.

_____

COMPUTE
Do the computation required to complete each task. Follow the steps listed on page 79.

COMMUNICATE
Write the measurement.

_____

5 Sam gives you a rod to measure. The gauge is shown here. Read the gauge. Write the measurement.

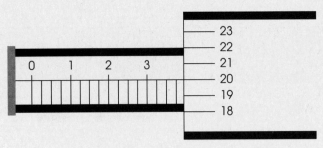

COMPREHEND
In the space below, write what you are to do.

_____

Write the steps you should follow to solve the problem.

_____

_____

_____

_____

Write the information you need to solve the problem.

_____

Write the system of measurement.

_____

COMPUTE
Do the computation required to complete each task. Follow the steps listed on page 79.

COMMUNICATE
Write the measurement.

_____

6

MATHEMATICS AND HUMAN RESOURCES

Part 6 covers some human resources you may deal with on the job. There is one job situation in this part of the book: producing graphs. The job situation has to do with people—the customers of A&P Auto. The job situation requires you to do several math tasks. Use the 3 Cs to complete each math task.

❏ JOB SITUATION 1   GRAPHS

When you work in the A&P Auto center office, you sometimes help Dominick and Conchita prepare reports. They need to know how the shop is doing. How much business did they do? What caused customers to pick this shop? How much did customers pay? How many tires did Dominick sell? Often they want to see this information in graphs.

You draw the graphs. You use graph paper, a compass, a protractor, a ruler, and pencil. The graphs you make look like the ones shown in Figures 6-1, 6-2, and 6-3.

Figure 6-1 A Bar Graph

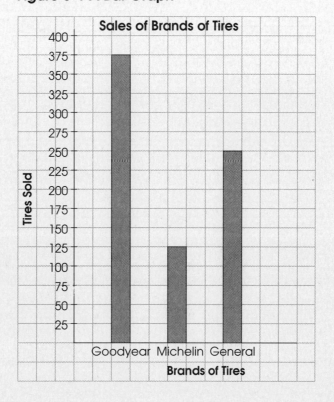

Figure 6-2 A Line Graph

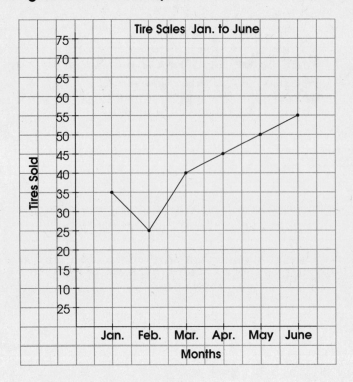

Figure 6-3 A Circle Graph

Sources of Sales

When you are asked to make a graph, you need to know what type of graph to make. You have a choice of three types of graphs: bar graphs, line graphs, and circle graphs. Figures 6-1, 6-2, and 6-3 show all three types. Each graph type is good for showing a different kind of information.

Bar graphs allow you to show and compare different numbers. The bar graph in Figure 6-1 shows how many customers picked each brand of tire. This information helps Dominick decide what brands to order. He can be sure of having the tires his customers want.

Line graphs allow you to show numbers and also to show changes over time. The line graph in Figure 6-2 shows sales over a period of six months. You can see what the sales were each month. You can also see when sales went up or down. Dominick uses this information to see when people are most likely to buy tires.

Circle graphs allow you to compare parts of a whole. The circle graph in Figure 6-3 shows reasons that customers gave for choosing to shop at A&P Auto. The whole circle stands for the total number of customers. Each wedge of the circle stands for customers who gave that reason.

Once you have picked the type of graph to draw, you need to make some other decisions. You have to decide what labels to give the graph, what scale to use, and how to draw the lines, wedges, or bars. Then you can make the graph.

TASK 1—CREATING A BAR GRAPH

Suppose that Dominick asked you to prepare a bar graph showing how many tires of each brand he has sold. The graph will look like the bar graph in Figure 6-1. Here are the steps you would follow to create the graph.

1. Draw a table to show the information.

Brand	Number of Tires Sold
Goodyear	375
Michelin	125
General	250

2. Choose a label for each **axis** of the graph. The **axes** of a graph are the vertical and horizontal lines. One axis will show the tire brands. One will show numerical data—the number of tires sold.

Label the vertical axis "Tires Sold." Label the horizontal axis "Brands of Tires."

3. Decide on the scale for the numerical axis. A scale should include the lowest to the highest number you need to show. It can go from just below to just above these numbers. The interval should be small enough to show the differences between the data but large enough so that you don't crowd the scale with extra numbers.

For the graph in Figure 6-1, a scale going from 0 to 400 works. The interval can be 25 tires.

4. Write information on the other axis. Pick a spot for each of the tire brands.

5. Draw bars to show each number in your table. For example, the bar for Goodyear goes to a point next to the "375" on the axis.

6. Write a title for the graph—"Sales of Brands of Tires."

To sum up:

1. Create a table for the data.
2. Choose a label for each axis.

3. Create a scale for the numerical axis.

4. Write the information on the other axis.

5. Draw a bar for each pair of numbers in your table.

6. Title the graph.

USING THE 3 CS
Dominick gives you these sales figures. Draw a bar graph to show the total sales for each brand of tire.

Goodyear: 300 tires   Michelin: 175 tires   General: 275 tires

COMPREHEND
In the space below, write what you are to do.

> *Draw a graph to show sales for each brand.*

Write the steps you should follow to solve the problem.

> *1. Create a table for the data.*
>
> *2. Choose a label for each axis.*
>
> *3. Create a scale for the numerical axis.*
>
> *4. Write the information on the other axis.*
>
> *5. Draw a bar for each pair of numbers in your table.*
>
> *6. Title the graph.*

Write the information you need to solve the problem.

> *Goodyear: 300 tires Michelin: 175 tires General: 275 tires*

Write the system of measurement.

> *Tires*

COMPUTE
Do the computation required to complete each task.
Follow the steps listed on pages 87–88.

> *1. Table:*

Brand	Number of Tires Sold
Goodyear	375
Michelin	125
General	250

> *2. Labels: "Tires Sold," "Brands of Tires"*
>
> *3. Scale: 0 to 400 Interval: 25*
>
> *4. Bars: See the graph below.*
>
> *5. Title: "Sales of Brands of Tires"*

COMMUNICATE
Draw the graph.

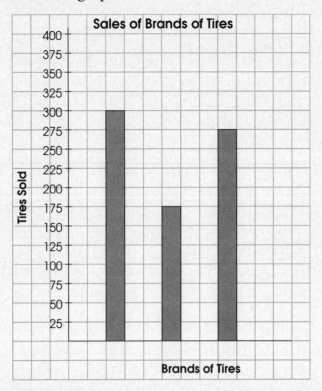

PRACTICE

1 Dominick gives you these sales figures. Draw a bar graph to show the total sales for each brand of tire.

Goodyear: 275 tires   Michelin: 100 tires   General: 200 tires

COMPREHEND
In the space below, write what you are to do.

_____

Write the steps you should follow to solve the problem.

_____

_____

_____

_____

_____

Write the information you need to solve the problem.

_____

_____

Write the system of measurement.

COMPUTE
Do the computation required to complete each task. Follow the steps on pages 87–88.

Brand	Number of Tires Sold

COMMUNICATE
Draw the graph. Use the graph paper shown here.

2 Dominick gives you these sales figures. Draw a bar graph to show the total sales for each brand of tire.

Goodyear: 375 tires   Michelin: 250 tires   General: 325 tires

COMPREHEND
In the space below, write what you are to do.

_____

Write the steps you should follow to solve the problem.

_____

_____

_____

_____

_____

_____

Write the information you need to solve the problem.

_____

_____

Write the system of measurement.

_____

COMPUTE
Do the computation required to complete each task. Follow the steps on pages 87–88.

Brand	Number of Tires Sold

COMMUNICATE
Draw the graph. Use the graph paper shown here.

TASK 2—CREATING A LINE GRAPH

When you make a line graph to show information, you follow steps very similar to those for making a bar graph.

1. Draw a table to show the information. Here is the table for the line graph shown in Figure 6-2.

Month	Tire Sales
Jan.	35
Feb.	25
Mar.	40
Apr.	45
May	50
June	55

2. Choose a label for each axis of the graph. One axis will show the months. One will show numerical data—the number of tires sold in each month.

Label the vertical axis "Tires Sold." Label the horizontal axis "Months."

3. Decide on the scale for numerical axis.

For the graph in Figure 6-2, a scale going from 0 to 75 works. The interval is 5.

4. Write information on the other axis. Pick a spot for each of the months.

5. Draw dots to show each number in your table. For example, the dot for January goes to a point even with the "35" on the axis. Connect the dots with lines.

6. Write a title for the graph—"Tire Sales Jan. to June."

To sum up:

1. Create a table for the data.
2. Choose a label for each axis.
3. Create a scale for the numerical axis.
4. Write the information on the other axis.
5. Draw a dot for each pair of numbers in your table. Connect the dots.
6. Title the graph.

USING THE 3 CS

Draw a line graph to show tire sales from January to June. Dominick gives you these figures: January—50, February—35, March—50, April—50, May—60, June—65.

COMPREHEND

In the space below, write what you are to do.

Draw a line graph to show tire sales each month.

Write the steps you should follow to solve the problem.

1. Create a table for the data.
2. Choose a label for each axis.
3. Create a scale for the numerical axis.
4. Write the information on the other axis.
5. Draw a dot for each pair of numbers in your table. Connect the dots.
6. Title the graph.

Write the information you need to solve the problem.

Sales were January—50, February—35, March—50, April—50, May—60, June—65.

Write the system of measurement.

Tires

COMPUTE
Do the computation required to complete each task. Follow the steps given on pages 92–93.

    1. Table:

Month	Tire Sales
Jan.	50
Feb.	35
Mar.	50
Apr.	50
May	60
June	65

    2. Labels: "Months," "Tires Sold"

    3. Scale: 0 to 70 Interval: 5

    4. Lines: See the graph below.

    5. Title: "Tire Sales Jan. to June"

COMMUNICATE
Draw the graph.

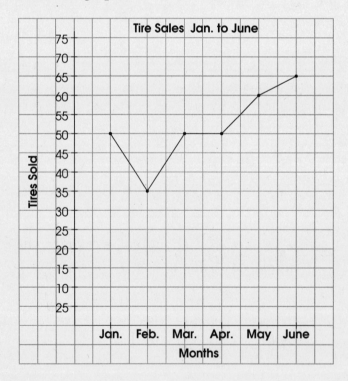

PRACTICE

3 Draw a line graph to show tire sales from January to June. Dominick gives you these figures: January—45, February—35, March—40, April—45, May—60, June—65.

COMPREHEND

In the space below, write what you are to do.

_____

Write the steps you should follow to solve the problem.

_____

_____

_____

_____

_____

_____

_____

Write the information you need to solve the problem.

_____

_____

Write the system of measurement.

_____

COMPUTE

Do the computation required to complete each task. Follow the steps given on pages 92–93.

Month	Tire Sales

COMMUNICATE
Draw the graph.

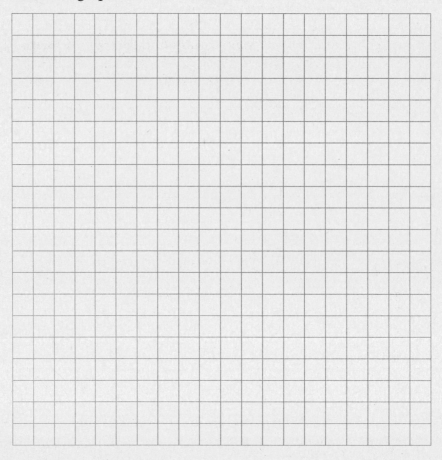

4 Draw a line graph to show tire sales from July to December. Dominick gives you these figures: July—65, August—60, September 65, October—50, November 45, December—40.

COMPREHEND
In the space below, write what you are to do.

_____

Write the steps you should follow to solve the problem.

_____

_____

_____

_____

_____

_____

Write the information you need to solve the problem.

_____

_____

Write the system of measurement.

COMPUTE

Do the computation required to complete each task. Follow the steps given on pages 92–93.

Month	Tire Sales

COMMUNICATE

Draw the graph.

TASK 3—MAKING CIRCLE GRAPHS

Circle graphs show the relationships between parts of a whole. The circle stands for the whole amount. The whole may be the total number of customers, as in the circle graph in Figure 6-3. Each section, or wedge, of the circle stands for one of the parts of the whole. Each wedge in Figure 6-3 is one of the reasons customers came A&P Auto. Each wedge is a percentage of the whole. The whole is 100 percent. The larger the percentage of the whole, the larger the wedge.

Here are the steps you would follow to create the circle graph shown in Figure 6-3.

1. Make a table showing the information. Here is the table for the circle graph in Figure 6-3.

Reason For Choosing A&P Auto?	Number of Customers Giving this Reason
Repeat Customer: Bought from A&P Before	300
Saw Ad in "Yellow Pages"	250
Recommendation From Friend	175
Saw Newspaper Ad	275

2. Find the percent of the whole that each item stands for.
   Repeat Customer 30%
   Ad in "Yellow Pages" 25%
   Recommendation 17.5%
   Newspaper Ad 27.5%

3. Find the number of degrees that each percent stands for. (There are 360° in a circle.)
   Repeat Customer: 108°
   Ad in "Yellow Pages": 90°
   Recommendation: 63°
   Newspaper Ad: 99°

4. Draw a circle with a compass.
   Use a protractor to draw each angle.

- Draw a line from the center of the circle to one side.

- Put the center of the protractor on the center of the circle and the 0° mark on the protractor on the line.

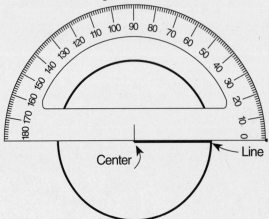

- To draw an angle of 36°, put a dot on the paper at the 36° mark on the protractor. Line up a ruler from the dot at the center of the circle to the dot. Draw a line from the center of the circle to the edge of the circle.

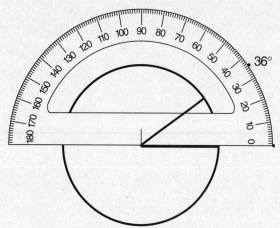

- To draw the next angle, place the center of the protractor at the center of the circle, but this time put the 0° mark on the line you just drew. (Turn the paper as necessary.) Be sure to label each angle.

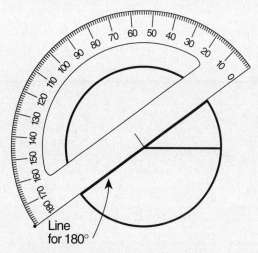

Line for 180°

5. Write a title for the graph.

To sum up:

1. Create a table for the data.
2. Find the percent of the whole for each item.
3. Find the number of degrees for each item.
4. Draw a circle and draw an angle for each item of data. Label it.
5. Title the graph.

USING THE 3 CS

You had 1000 people fill out a questionnaire. They told why they decided to buy tires from A&P Auto. Here are the results of the survey:

Reasons: Good Prices—500 people, Liked the Brands—275 people, Shop Is in a Good Location—225

Draw a circle graph to show this data.

COMPREHEND
In the space below, write what you are to do.

Make a circle graph to show why people bought from A&P.

Write the steps you should follow to solve the problem.

1. Create a table for the data.
2. Find the percent of the whole for each item.
3. Find the number of degrees for each item.
4. Draw a circle and draw an angle for each item of data. Label it.
5. Title the graph.

Write the information you need to solve the problem.

Total of 1000 people.

Reasons: Good Prices—500 people, Liked the Brands—275 people, Shop Is in a Good Location—225

Write the system of measurement.

People

COMPUTE
Do the computation required to complete each task.
Use the instructions given on pages 98–99.

1. Table:

2. Percentages:

"Prices" chosen by 500 out of 1000 people. What percent is 500 of 1000?

$$n \times 1000 = 500$$
$$\div 1000 \quad \div 1000$$
$$n = 0.50 = 50\%$$

"Brands" chosen by 275 out of 1000. What percent is 275 of 1000?
$$n \times 1000 = 275$$
$$n = 0.275 = 27.5\%$$

"Location" chosen by 225 out of 1000. What percent is 225 of 1000?
$$n \times 1000 = 225$$
$$n = 0.225 = 22.5\%$$

3. Angles:
   "Prices" is 50%. The angle is 50% of 360°.
      $$360° \times 0.50 = 180°$$
   "Brands" is 27.5%. Angle is 27.5% of 360°.
   $$360° \times 0.275\% = 99°$$
   "Location" is 22.5%. Angle is 22.5% of 360°.
      $$360° \times 0.225 = 81°$$

4. Draw a circle and draw an angle for each item of data. Label each wedge.

5. Title: "Reasons for Buying at A&P Auto."

COMMUNICATE
Make the graph.

**Reasons for Buying
at A&P Auto**

PRACTICE

5 You asked 1000 people why they decided to buy tires from A&P Auto. Here are the results of the survey:

Reasons: Good Prices—500 people, Liked the Brands—200 people, Shop Is in a Good Location—300

Draw a circle graph to show this data.

COMPREHEND
In the space below, write what you are to do.

_____

_____

Write the steps you should follow to solve the problem.

_____

_____

_____

_____

_____

Write the information you need to solve the problem.

_____

_____

Write the system of measurement.

_____

COMPUTE

Do the computation required to complete each task. Use the instructions given on pages 98–99.

_____

_____

_____

_____

_____

_____

_____

_____

_____

COMMUNICATE
Make the graph.

6 You asked 1000 people how happy they were with the service they got at A&P. Here are the results of the survey:

Very Satisfied—200 people, Satisfied—700 people, Not Satisfied—100 people

Draw a circle graph to show this data.

COMPREHEND

In the space below, write what you are to do.

_____

_____

Write the steps you should follow to solve the problem.

_____

_____

_____

_____

_____

Write the information you need to solve the problem.

_____

_____

Write the system of measurement.

_____

COMPUTE

Do the computation required to complete each task. Use the instructions given on pages 98–99.

_____

_____

_____

_____

_____

_____

_____

_____

COMMUNICATE
Make the graph.

MATHEMATICS REVIEW

❏ BASIC OPERATIONS

PLACE VALUE

The **digits** 0 to 9 are used to name any number. The **value** of a digit depends on its place in the number. For example, in each of these numbers, the digit 5 has a different value.

5	This 5 means 5 ones.
50	This 5 means 5 tens.
500	This 5 means 5 hundreds.
5,000	This 5 means 5 thousands.

To read any number, you need to know the value of each place. This chart shows **place value** up to hundred millions.

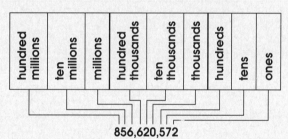

856,620,572

This number means:
8 hundred millions   + 5 ten millions   + 6 millions   +
6 hundred thousands + 2 ten thousands + 0 thousands +
5 hundreds   + 7 tens   + 2 ones

78,019

This means:
7 ten thousands, 8 thousands
0 hundreds, 1 ten, 9 ones

Long numbers are separated into sections by commas. To read a number, read the value of each of the sections.

(32),(621)

Read as: thirty-two thousand, six hundred twenty-one

(856),(620),(572)

Read as:
eight hundred fifty-six million, six hundred twenty thousand, five hundred seventy-two

78,019

Read as: seventy-eight thousand, nineteen

150,399

Read as: one hundred fifty thousand, three hundred ninety-nine

15,399

Read as: fifteen thousand, three hundred ninety-nine

ADDITION

When adding any set of numbers, first write the numbers with the ones in the ones column, the tens in the tens column, and so on.

Example: 912 + 36 $\longrightarrow$

$$\begin{array}{r} 9\ 1\ 2 \\ +\quad 3\ 6 \\ \hline \end{array}$$

Then follow these steps:

Step 1:   Add the ones and write the answer in the ones column.

2 + 6 = 8

$$\begin{array}{r} 912 \\ +\ 36 \\ \hline 8 \end{array}$$

Step 2:   Add the tens and write the answer in the tens column.

1 + 3 = 4

$$\begin{array}{r} 912 \\ +\ 36 \\ \hline 48 \end{array}$$

Step 3:   Add the hundreds. Notice that in this case there is only one number in the hundreds column.

9 + 0 = 9

$$\begin{array}{r} 912 \\ +\ 36 \\ \hline 948 \end{array}$$

Write the 9 in the hundreds column. The answer is 948.

Here's an example that shows what to do when the numbers you get are too large for the column. See how you **regroup** when the number is too large.

647 + 598 $\longrightarrow$

$$\begin{array}{r} 647 \\ +598 \\ \hline \end{array}$$

Step 1:   Add the ones.

7 + 8 = 15

You can't write 15 in the ones column, so think of 15 as 1 ten and 5 ones. Write the 5 ones in the ones column. Write the 1 ten at the top of the tens column.

$$\begin{array}{r} 1\ \ \\ 647 \\ +598 \\ \hline 5 \end{array}$$

Step 2: Add the tens. Don't forget the 1 you wrote at the top of the tens column.

$1 + 4 + 9 = 14$

You can't write 14 in the tens column, so think of the 14 as 1 hundred and 4 tens. Write the 4 in the tens column and write the 1 hundred in the hundreds column.

$$\begin{array}{r} 1\ 1 \\ 647 \\ +598 \\ \hline 45 \end{array}$$

Step 3: Add the hundreds.

$1 + 6 + 5 = 12$

Write the 2 in the hundreds column and write the 1 in the thousands column. The answer is 1,245.

$$\begin{array}{r} 1\ 1 \\ 647 \\ +598 \\ \hline 1{,}245 \end{array}$$

To add larger numbers, follow the same steps. Write the number with the ones lined up, the tens lined up, and so on. Then add the ones, the tens, the hundreds, the thousands, and so on. Remember to write any number you have to regroup in the next column.

$$\begin{array}{rrrr} & 1\ 2\ 1 & & \\ & 34{,}921 & 1\ 2\ 2 & \\ & 15{,}780 & 126{,}781 & 1\ 1\ 1\quad\ 1 \\ 1 & 5{,}711 & 340{,}990 & 16{,}782{,}229 \\ 4{,}573 & +23{,}002 & +\ \ 7{,}631 & +\ 5{,}757{,}001 \\ +6{,}118 & \hline & \hline & \hline \\ \hline 10{,}691 & 79{,}414 & 475{,}402 & 22{,}539{,}230 \end{array}$$

SUBTRACTION

The **inverse** of addition is subtraction. Subtraction undoes addition. Every addition fact can be undone in two ways as subtraction facts.

Addition: $7 + 2 = 9$
Subtraction: $9 - 2 = 7$
 $9 - 7 = 2$

You can use what you know about addition to help you solve subtraction problems. Here's an example:

$$15 - 7 = ?$$

Ask "What number added to 7 gives 15?" The answer is 8.

$$15 - 7 = 8$$

Addition can also be used to check the answer in subtraction.

$$\begin{array}{r} 23 \\ -\ 7 \\ \hline 16 \end{array} \qquad \text{Check:} \qquad \begin{array}{r} 16 \\ +\ 7 \\ \hline 23\ \checkmark \end{array}$$

When subtracting, first write the numbers with the ones in the ones column, the tens in the tens column, and so on.

Example: 176 − 31 ⟶   176
                          − 31

Then follow these steps:

Step 1: Subtract the ones and write the answer in the ones column.

6 − 1 = 5

```
  176
−  31
    5
```

Step 2: Subtract the tens and write the answer in the tens column.

7 − 3 = 4

```
  176
−  31
   45
```

Step 3: Subtract the hundreds. In this case, there is only one number in the hundreds column. Write the 1. The answer is 145.

```
  176
−  31
  145
```

Step 4: To check your answer, you can add from the bottom.

145 + 31 = 176

```
  176  ⟶  145
−  31     + 31
  145     176√
```

Here's an example in which some of the digits you are subtracting are greater than the digits you are subtracting from. You need to regroup in order to subtract in that column.

1,264 − 627 ⟶   1264
                  − 627

Step 1: Subtract the ones. You can't subtract 7 from 4, so you need to use a number from the tens column. Regroup 6 tens and 4 ones as 5 tens and 14 ones. Write the 5 above the tens column as a reminder and write the 14 in the ones column. Now subtract 7 from 14 and write the answer in the ones column.

```
     5 14
  1 2 6̸ 4̸
−    6 2 7
        7
```

Step 2: Subtract the tens. Don't forget that the number in the tens column is now 5.

5 − 2 = 3

Write the answer in the tens column.

```
     5 14
  1 2 6̸ 4̸
−    6 2 7
       3 7
```

Step 3: Subtract the hundreds. You can't subtract 6 from 2, so you need to use a number from the thousands column. Regroup 1 thousand and 2 hundreds as 12 hundreds. Cross out the 1 thousand and write the 12 in the hundreds column. Subtract 6 from 12 and write the answer in the hundreds column. There are no thousands to subtract. The answer is 637.

$$
\begin{array}{r}
{\scriptstyle 12\ 5\ 14} \\
1\,2\,6\,4 \\
-\quad 6\,2\,7 \\
\hline
6\,3\,7
\end{array}
\quad\longrightarrow\quad
\begin{array}{r}
{\scriptstyle 1} \\
6\,3\,7 \\
+\,6\,2\,7 \\
\hline
1\,2\,6\,4\,\checkmark
\end{array}
$$

Check the answer by adding from the bottom.

$637 + 627 = 1,264$

To subtract larger numbers, follow the same steps. Write the number with ones lined up, tens lined up, and so on. Then subtract the ones, the tens, the hundreds, the thousands, and so on. Remember to cross out and rewrite any numbers you have to regroup.

$$
\begin{array}{r}
{\scriptstyle 8\ 13} \\
3\,4,9\,3\,9 \\
-\quad 2,8\,6\,2 \\
\hline
3\,2,0\,7\,7
\end{array}
\qquad
\begin{array}{r}
{\scriptstyle 8\ 12\quad 0\ 10} \\
4\,9\,2,6\,1\,0 \\
-\,1\,0\,9,3\,0\,9 \\
\hline
3\,8\,3,3\,0\,1
\end{array}
\qquad
\begin{array}{r}
{\scriptstyle\ \ \ 10} \\
{\scriptstyle 4\ \ 0\ 13} \\
1\,5,1\,3\,8 \\
-\quad 2,5\,5\,7 \\
\hline
1\,2,5\,8\,1
\end{array}
$$

MULTIPLICATION

You can think about multiplication as repeated addition. The multiplication fact 6×5 means 6 groups of 5.

This is the same as adding the 6 groups of 5.

$$5 + 5 + 5 + 5 + 5 + 5 = 30$$
$$①\quad ②\quad ③\quad ④\quad ⑤\quad ⑥$$

The answer in multiplication is called the **product.** The product of 6 and 5 is 30.

You are probably familiar with the multiplication facts—$2 \times 2 = 4$; $2 \times 3 = 6$; $2 \times 4 = 8$.... If you are unsure of any of these facts, it would be a good idea to make flash cards for yourself and memorize the facts. They are the basis of any multiplication that you do.

When you multiply, first write the numbers with the ones in the ones column, the tens in the tens column, and so on. This helps you keep track of the value of each digit as you multiply.

$$357 \times 4 \longrightarrow \begin{array}{r} 357 \\ \times\ 4 \\ \hline \end{array}$$

Step 1: Always multiply the **factor** on the bottom times the factor on the top. Start by multiplying the ones by the ones.

$7 \times 4 = 28$

You can't write the product 28 in the ones column, so regroup as 2 tens and 8 ones. Write the 8 and write the 2 above the tens column.

$$\begin{array}{r} 2 \\ 357 \longleftarrow \text{factor} \\ \times \quad 4 \longleftarrow \text{factor} \\ \hline 8 \end{array}$$

Step 2: Multiply the tens by the ones.

$5 \times 4 = 20$

Add the 2 that you wrote in the tens column. You can't write 22 in the tens column. Regroup.

$$\begin{array}{r} 2\,2 \\ 357 \\ \times \quad 4 \\ \hline 28 \end{array}$$

Step 3: Multiply the hundreds by the ones.

$3 \times 4 = 12$

Add the 2. Write the 14. The answer is 1,428.

$$\begin{array}{r} 2\,2 \\ 357 \\ \times \quad 4 \\ \hline 1,428 \end{array}$$

Here's how to multiply by a two-digit number.

Step 1: Multiply the ones by the ones.

$4 \times 6 = 24$

You can't write 24 in the ones column, so regroup as 2 tens and 4 ones. Write the 4 and write the 2 above the tens column.

$$\begin{array}{r} 2 \\ 54 \\ \times 36 \\ \hline 4 \end{array}$$

Step 2: Multiply the tens by the ones and add the 2.

$5 \times 6 = 30 + 2 = 32$

Write the 32.

$$\begin{array}{r} 2 \\ 54 \\ \times 36 \\ \hline 324 \end{array}$$

Step 3: Now start multiplying by the tens. Multiply ones by tens.

$4 \times 3 = 12$

The 3 in the tens place stands for 30, so it's really $4 \times 30 = 120$. Regroup. Write the 20. Write the 1 above the next column.

$$\begin{array}{r} 1 \\ 54 \\ \times 36 \\ \hline 324 \\ 20 \end{array}$$

Step 4: Multiply tens by tens and add the 1.

$5 \times 3 = 15 + 1 = 16$

Write the 16.

$$\begin{array}{r} 1 \\ 54 \\ \times 36 \\ \hline 324 \\ 1,620 \end{array}$$

Step 5: Now add the products.

$$324 \times 1620 = 1944$$

$$
\begin{array}{r}
54 \\
\times 36 \\
\hline
324 \\
1\ 620 \\
\hline
1,944
\end{array}
$$

When you multiply, be careful to write each product in the right column or columns. Add the products to get the final answer.

$$
\begin{array}{r}
239 \\
\times 23 \\
\hline
717 \\
+4\ 780 \\
\hline
5,497
\end{array}
\qquad
\begin{array}{r}
704 \\
\times 123 \\
\hline
2\ 112 \\
14\ 080 \\
+70\ 400 \\
\hline
86,592
\end{array}
$$

Write the tens here.

Write 4 × 100 here.

DIVISION

You can use the multiplication facts that you know to figure out division facts.

$$3 \times 5 = 15 \longrightarrow 15 \div 5 = 3 \text{ and } 15 \div 3 = 5$$

The answer in division is called the **quotient.** Here's how to find the quotient when you divide by a one-digit number.

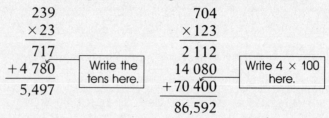

Step 1: Divide the hundreds by 6.

$6 \div 6 = 1$

Write the 1. Multiply 1 × 6.
Write the 6. Subtract.

$$
\begin{array}{r}
1 \\
6\overline{)672} \\
(1 \times 6 =)\quad 6 \\
\hline
0
\end{array}
$$

Step 2: Bring down the tens. Divide 7 by 6. The closest you can get to the answer is 1. Write the 1. Multiply 1 × 6. Write the 6. Subtract.

$$
\begin{array}{r}
11 \\
6\overline{)672} \\
6 \\
\hline
7 \\
(1 \times 6 =)\quad 6 \\
\hline
1
\end{array}
$$

Step 3: Bring down the ones. Divide 12 by 6. Write the 2. Multiply 2 × 6. Subtract. There is no **remainder.**

$$
\begin{array}{r}
112 \\
6\overline{)672} \\
6 \\
\hline
7 \\
6 \\
\hline
12 \\
(2 \times 6 =)\quad 12 \\
\hline
0
\end{array}
$$

Here's another example.

Step 1: Try to divide the thousands. You can't divide 3 by 8. Divide the hundreds.

$31 \div 8$

The closest you can get to the answer is 3. Multiply. Subtract.

$$
\begin{array}{r}
3 \\
8\overline{)3153} \\
(3 \times 8 =)\quad 24 \\
\hline
7
\end{array}
$$

Step 2:	Bring down the tens. Divide 75 by 8. Write 9. Multiply. Subtract.		

$$\begin{array}{r} 39 \\ 8\overline{)3153} \\ 24 \\ \hline 75 \\ (9 \times 8 =) \quad 72 \\ \hline 3 \end{array}$$

Step 3: Bring down the ones. Divide 33 by 8. Write 4. Multiply. Subtract. There is 1 one left over. Write it as a remainder.

Quotient: 394 *R1*

$$\begin{array}{r} 394\ R1 \\ 8\overline{)3153} \\ 24 \\ \hline 75 \\ 72 \\ \hline 33 \\ (4 \times 8 =) \quad 32 \\ \hline 1 \end{array}$$

Here's an example in which there is a zero in the quotient.

Step 1: Divide the hundreds. Multiply. Subtract.

$$\begin{array}{r} 1 \\ 8\overline{)864} \\ (1 \times 8 =) \quad 8 \\ \hline 0 \end{array}$$

Step 2: Bring down the tens. You can't divide 6 by 8. Write a 0 in the quotient. Multiply. Subtract.

$$\begin{array}{r} 10 \\ 8\overline{)864} \\ 8 \\ \hline 6 \\ (0 \times 8 =) \quad 0 \\ \hline 6 \end{array}$$

Step 3: Bring down the ones. Divide. Multiply. Subtract. There is no remainder.

$$\begin{array}{r} 108 \\ 8\overline{)864} \\ 8 \\ \hline 6 \\ 0 \\ \hline 64 \\ (8 \times 8 =) \quad 64 \\ \hline 0 \end{array}$$

You follow the same steps when you divide by a two-digit number. Use the division facts to help you estimate.

Step 1: Try to divide the hundreds. You can't divide 3 by 62. Try the tens. You can't divide 32 by 62. Divide the ones.

$321 \div 62$

A good estimate is 5 ($5 \times 6 = 30$, so $5 \times 60 = 300$).

$$\begin{array}{r} 5 \\ 62\overline{)321} \end{array}$$

Step 2: Multiply 5×62. Subtract. Write the remainder.

$$\begin{array}{r} 5\ R11 \\ 62\overline{)321} \\ 310 \\ \hline 11 \end{array}$$

Here's another example.

Step 1: You can't divide the hundreds. Divide the tens.

81 ÷ 24

Try 4 (4 × 2 = 8).

Multiply 4 × 24. The answer is too large.

$$\begin{array}{r} 4 \\ 24\overline{)813} \\ (4 \times 24 =)\quad 96 \end{array}$$

Step 2: Try a lower estimate—3. Multiply 3 × 24. Subtract.

$$\begin{array}{r} 3 \\ 24\overline{)813} \\ (3 \times 24 =)\quad \underline{72} \\ 9 \end{array}$$

Step 3: Bring down the ones. Divide. Multiply. Subtract. Write the remainder.

$$\begin{array}{r} 33\ R21 \\ 24\overline{)813} \\ \underline{72} \\ 93 \\ \underline{72} \\ 21 \end{array}$$

Divide greater numbers in the same way.

Step 1: You can't divide the thousands. Divide the hundreds.

46 ÷ 34

Write the estimate. Multiply. Subtract.

$$\begin{array}{r} 1 \\ 34\overline{)4602} \\ \underline{34} \\ 12 \end{array}$$

Step 2: Bring down the tens. Divide.

120 ÷ 34

Try 4. Too large. Try 3. Multiply. Subtract.

$$\begin{array}{r} 13 \\ 34\overline{)4602} \\ \underline{34} \\ 120 \\ \underline{102} \\ 18 \end{array}$$

Step 3: Bring down the ones.

182 ÷ 34

Try 5. Multiply. Subtract. Write the remainder.

$$\begin{array}{r} 135\ R12 \\ 34\overline{)4602} \\ \underline{34} \\ 120 \\ \underline{102} \\ 182 \\ \underline{170} \\ 12 \end{array}$$

❏ ORDER OF OPERATIONS

Some math problems require doing more than one kind of operation. For example, this problem requires addition, subtraction, and multiplication.

$$[4 + (10 - 7)] \times 3$$

To solve problems like this, you need to know some basic principles.

1. Do multiplication and division first. Then do addition and subtraction. Work from the left to the right.

Divide.

$2 + \boxed{16 \div 4} =$

$2 + \quad 4 \quad = 6$

Then add.

Multiply.

$3 + \boxed{8 \times 4} =$

$3 + \quad 32 \quad = 35$

Then add.

$30 - \boxed{8 \times 2} + 1 =$

$\boxed{30 - \quad 16} \quad + 1 =$

$14 \qquad + 1 = 15$

$\boxed{2 \times 6} + \boxed{24 \div 8} =$

$12 \quad + \quad 3 \quad = 15$

2. When you see symbols such as [ ] or ( ), do the operations inside the symbols first.

Examples: $(3 + 8) \times 2 =$

$11 \quad \times 2 = 22$

$25 \div (3 + 2) =$

$25 \div \quad 5 \quad = 5$

3. Do the operations inside the innermost symbols first.

$[4 + (10 - 7)] \times 3 =$

$[4 + \quad 3] \quad \times 3 =$

$7 \qquad \times 3 = 21$

❏ ROUNDING

Often it isn't necessary to work with exact numbers. Instead, you can make estimates using round numbers.

❏ For example, if 48 people will attend a meeting, round the number up to 50 and set up 50 chairs.

❏ If you are going to buy 11 boxes of paper costing $19.99 each, use the round number $20 for the cost and multiply by 10 to get an estimate of the total cost.

To round these numbers to tens, look at the ones position. Increase the tens digit if the ones digit is 5 or greater.

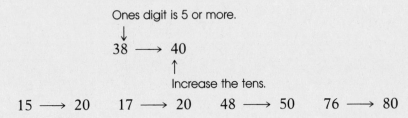

Ones digit is 5 or more.

$38 \longrightarrow 40$

Increase the tens.

$15 \longrightarrow 20 \qquad 17 \longrightarrow 20 \qquad 48 \longrightarrow 50 \qquad 76 \longrightarrow 80$

Leave the number the same if the digit is less than 5.

Ones digit is less than 5.
↓
34 ⟶ 30
↑
Stays the same.

14 ⟶ 10   22 ⟶ 20   41 ⟶ 40   75 ⟶ 80

To round to hundreds, look at the tens digit. Is it 5 or greater or less than 5?

Tens digit is 5 or more.   Tens digit is less than 5.
↓   ↓
161 ⟶ 200   329 ⟶ 300
↑   ↑
Round up.   Stays the same.

When you round, look to the digit to the right of the place you are rounding to.

Round to thousands: Round to ten thousands:
4,|1|89 ⟶ 4000   5|9|,012 ⟶ 60,000
   ↑   ↑
Stays the same.   Round up.

Round to hundred thousands: Round to millions:
4|2|9,228 ⟶ 400,000   3,|6|21,008 ⟶ 4,000,000

❏ FRACTIONS

FRACTIONS AND MIXED NUMBERS

Fractions are parts of a whole. This pizza is cut into 8 equal pieces. Each piece is ⅛ (one-eighth) of the whole. If there are five pieces left, then ⅝ (five-eighths) of the pizza is left.

 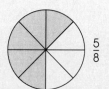

The top number of a fraction is called the **numerator.** The bottom number of a fraction is called the **denominator.** Each number has a meaning:

$\dfrac{3}{4}$ numerator—shaded parts
   denominator—how many in all

$\dfrac{1}{5}$ shaded part
   how many in all

The fraction ⅝ refers to 5 parts out of a total of 8 parts: 5 pizza slices out of 8, for example.

A **mixed number** includes a whole number and a fraction. For example, 1¾ is a mixed number.

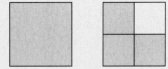

You can write this mixed number as a fraction. There are ⁴⁄₄ in the square on the left. There are ¾ in the square on the right.

$$1\frac{3}{4} = \frac{7}{4} \quad \begin{array}{l}\text{shaded parts} \\ \text{how many in each whole}\end{array}$$

To change a mixed number to a fraction, follow these steps:

$$2\frac{1}{2} = ?$$

Step 1: Write the denominator.
Multiply the whole number
times the denominator.

$2 \times 2 = 4$

$$2 \times \begin{array}{c}4\\ \nearrow \end{array} \frac{1}{2} = \frac{}{2}$$

Step 2: Add the numerator.
$4 + 1 = 5$

$$4 + \downarrow$$
$$2 \quad \frac{1}{2} = \frac{5}{2}$$

To change a fraction to a mixed number, follow these steps:

Step 1: Divide the numerator by the denominator. Write the quotient as the whole number part of the mixed number.

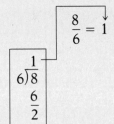

$$\frac{8}{6} = 1$$

$$\begin{array}{r} 1 \\ 6\overline{)8} \\ \underline{6} \\ 2 \end{array}$$

Step 2: Write the remainder as the numerator of the fraction part of the mixed number.

$$\frac{8}{6} = 1\frac{2}{6}$$

$$\begin{array}{r} 1 \\ 6\overline{)8} \\ \underline{6} \\ 2 \end{array}$$

EQUIVALENT FRACTIONS

There are many ways to name the same fraction, for example, ½, ⁴⁄₈, ³⁄₆ all name the same fraction. They are **equivalent fractions.**

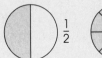

To find equivalent fractions, multiply or divide both the numerator and denominator by the same number.

$$\frac{1\ (\times\ 2)}{2\ (\times\ 2)} = \frac{2}{4}$$ The fractions ½ and ¼ are equivalent.

$$\frac{6\ (\div\ 3)}{9\ (\div\ 3)} = \frac{2}{3}$$ The fractions 6/9 and ⅔ are equivalent.

LOWEST TERMS

Fractions are **reduced to lowest terms** when you can't divide the numerator and denominator by any number other than 1 (which doesn't change the number).

$$\frac{5\ (\div\ 5)}{10\ (\div\ 5)} = \frac{1}{2}$$ This fraction is reduced to lowest terms. You can't divide both 1 and 2 by any number (except 1).

$$\frac{4\ (\div\ 2)}{8\ (\div\ 2)} = \frac{2}{4}$$ This fraction is not reduced to lowest terms. You can still divide 2 and 4 by 2. The lowest term is ½.

ADDING FRACTIONS

To add fractions, follow these steps:

Step 1: Add the numerators.

$$\frac{3}{8} + \frac{1}{8} \rightarrow \frac{4}{=}$$

Step 2: Write the denominator. It doesn't change.

$$\frac{3}{8} + \frac{1}{8} = \frac{4}{8}$$

Step 3: Reduce the fraction to lowest terms.

$$\frac{3}{8} + \frac{1}{8} = \frac{4\ (\div\ 4)}{8\ (\div\ 4)} = \frac{1}{2}$$

To add fractions with unlike denominators:

Step 1: Find a **common denominator** for the fractions. (This is a number that you can divide by both denominators.) One way to find a common denominator is to multiply the denominators.

$$\frac{2}{3} = \frac{}{12}$$

$$+\frac{3}{4} = \frac{}{12}$$

$3 \times 4 = 12$

You can divide 12 by both 3 and 4.

Step 2: Write equivalent fractions. Multiply the numerator by the same number that you used to find a common denominator.

$$\frac{2\ (\times\ 4)}{3\ (\times\ 4)} \frac{8}{12}$$

$$+\frac{3\ (\times\ 3)}{4\ (\times\ 3)} \frac{9}{12}$$

Step 3: Add. Write the answer in lowest terms.

$$\frac{2}{3} = \frac{8}{12}$$

$$+\frac{3}{4} = \frac{9}{12}$$

$$\frac{17}{12} = 1\frac{5}{12}$$

To add mixed numbers:

Step 1: Write equivalent fractions with common denominators.

$$2\frac{5}{8} \rightarrow 2\frac{15}{24}$$

$$+7\frac{5}{6} \rightarrow +7\frac{20}{24}$$

Step 2: Add the fractions. If the fraction can be changed to a mixed number, change it. Write the whole number part above the whole numbers.

$$2\frac{5}{8} \rightarrow \overset{1}{2}\frac{15}{24}$$

$$+7\frac{5}{6} \rightarrow +7\frac{20}{24}$$

$$\frac{\cancel{35}}{24}$$

$$\frac{11}{24}$$

$$24\overline{)35} \quad \frac{24}{11} \quad 1$$

Step 3: Add the whole numbers. Write the answer in lowest terms.

$$2\frac{5}{8} \rightarrow \overset{1}{2}\frac{15}{24}$$

$$+7\frac{5}{6} \rightarrow +7\frac{20}{24}$$

$$10\frac{11}{24}$$

SUBTRACTING FRACTIONS

To subtract fractions, follow these steps:

Step 1: Subtract the numerators.

$$\frac{9}{16} - \frac{5}{16} \rightarrow \frac{4}{}$$

Step 2: Write the denominator. It doesn't change.

$$\frac{9}{16} - \frac{5}{16} = \frac{4}{16}$$

Step 3: Reduce the fraction to lowest terms.

$$\frac{9}{16} - \frac{5}{16} = \frac{4\,(\div 4)}{16\,(\div 4)} = \frac{1}{4}$$

To subtract fractions with unlike denominators:

Step 1: Write equivalent fractions with a common denominator.

$$\frac{3}{4} = \frac{12}{16}$$

$$-\frac{1}{8} = \frac{2}{16}$$

Step 2: Subtract the numerators. Write the denominator.

$$\frac{3}{4} = \frac{12}{16}$$

$$-\frac{1}{8} = \frac{2}{16}$$

$$\frac{10}{16}$$

Step 3: Write the answer in lowest terms.

$$\frac{3}{4} = \frac{12}{16}$$

$$-\frac{1}{8} = \frac{2}{16}$$

$$\frac{10\,(\div 2)}{16\,(\div 2)} = \frac{5}{8}$$

To subtract mixed numbers:

Step 1: Write equivalent fractions with common denominators.

$$3\frac{1}{3} \rightarrow 3\frac{2}{6}$$

$$-1\frac{5}{6} \rightarrow 1\frac{5}{6}$$

Think of 3 as $2\frac{6}{6}$.

Step 2: Subtract the fractions. You can't subtract the numerator 5 from 2. Regroup 3²⁄₆ into 2 + ⁶⁄₆ + ²⁄₆ then into 2⁸⁄₆. Subtract the numerators. Write the denominator.

$$3\frac{1}{3} \rightarrow 3\frac{2}{6} \rightarrow 2\frac{6}{6} + \frac{2}{6} \rightarrow 2\frac{8}{6}$$

$$-1\frac{5}{6} \rightarrow 1\frac{5}{6} \longrightarrow 1\frac{5}{6}$$

$$\frac{3}{6}$$

Step 3: Subtract the whole numbers. Write the answer in lowest terms.

$$3\frac{1}{3} \rightarrow 2\frac{8}{6}$$

$$-1\frac{5}{6} \rightarrow 1\frac{5}{6}$$

$$1\frac{3}{6} = 1\frac{1}{2}$$

MULTIPLYING FRACTIONS

To multiply fractions, follow these steps.

Step 1: Write the fractions horizontally, with the numerators lined up and the denominators lined up.

$$\frac{2}{5} \times \frac{5}{8} =$$

Step 2: Multiply the numerators.

$2 \times 5 = 10$

Multiply the denominators.

$5 \times 8 = 40$

$$\frac{2}{5} \times \frac{5}{8} \Rightarrow \frac{10}{40}$$

Step 3: Write the fraction in lowest terms.

$$\frac{2}{5} \times \frac{5}{8} = \frac{10\,(\div\,10)}{40\,(\div\,10)}\frac{1}{4}$$

Any whole number can be written as a fraction with 1 in the denominator.

$$3 = \frac{3}{1} \qquad 2 = \frac{2}{1} \qquad 6 = \frac{6}{1} \qquad 15 = \frac{15}{1} \qquad 7 = \frac{7}{1}$$

To multiply a whole number times a fraction:

Step 1: Write the whole number as a fraction.

$$6 = \frac{6}{1}$$

$$6 \times \frac{2}{5} = \frac{6}{1} \times \frac{2}{5}$$

Step 2: Multiply as usual.

$$6 \times \frac{2}{5} = \frac{6}{1} \times \frac{2}{5} \rightrightarrows \frac{12}{5}$$

Step 3: Change the fraction to a mixed number.

$$\frac{6}{1} \times \frac{2}{5} = \frac{12}{5} = 2\frac{2}{5}$$

$$5)\overline{12}$$
$$\underline{10}$$
$$2$$

To multiply mixed numbers:

Step 1: Write the mixed numbers as fractions.

$$1\frac{1}{4} \times 2\frac{1}{2} =$$
$$\downarrow \qquad \downarrow$$
$$\frac{5}{4} \times \frac{5}{2}$$

Step 2: Multiply.

$$\frac{5}{4} \times \frac{5}{2} = \frac{25}{8}$$

Step 3: Write a mixed number for the answer.

$$\frac{5}{4} \times \frac{5}{2} = \frac{25}{8} = 3\frac{1}{8}$$

$$8)\overline{25}$$
$$\underline{24}$$
$$1$$

DIVIDING FRACTIONS

When you divide fractions, turn the problem into a multiplication problem with fractions. Then multiply as usual.

Step 1: Invert the number you are dividing by. Write ⅓ as ³⁄₁.

$$\frac{5}{9} \div \frac{1}{3} =$$
$$\downarrow$$
$$\frac{3}{1}$$

Step 2: Multiply as usual.

$$\frac{5}{9} \div \frac{1}{3} =$$
$$\frac{5}{9} \times \frac{3}{1} = \frac{15}{9}$$

Step 3: Write the answer in lowest terms.

$$\frac{5}{9} \times \frac{3}{1} = \frac{15}{9} = 1\frac{6}{9} = 1\frac{2}{3}$$

(lowest terms)

To divide by a whole number or mixed number:

Step 1: If you have a whole number or a mixed number, write it as a fraction.

$2\frac{1}{4} = \frac{9}{4}$

$$\frac{7}{8} \div 2\frac{1}{4}$$
$$\downarrow$$
$$\frac{7}{8} \div \frac{9}{4}$$

Step 2: Invert the number you are dividing by.

$\frac{9}{4} \longrightarrow \frac{4}{9}$

Make the numerator into the denominator.
Make the denominator into the numerator.

$$\frac{7}{8} \div \frac{9}{4}$$
$$\downarrow$$
$$\frac{4}{9}$$

Step 3: Multiply. Write the answer in lowest terms.

$$\frac{7}{8} \times \frac{4}{9} = \frac{28\,(\div 4)}{72\,(\div 4)} = \frac{7}{18}$$

❏ DECIMALS

PLACE VALUE IN DECIMALS

Fractions where the denominator is 10 or a multiple of 10—100, 1,000, 10,000 . . . —can be written as **decimals.**

Ordinary Fraction	Decimal
$\frac{6}{10}$	0.6

Mixed Number	Decimal
$2\frac{33}{100}$	2.33

Examples:

$$\frac{3}{10} = 0.3 \qquad \frac{4}{100} = 0.04 \qquad \frac{8}{1000} = 0.008$$

This chart shows place value of the decimal numbers. The decimal point separates the whole numbers from the decimals.

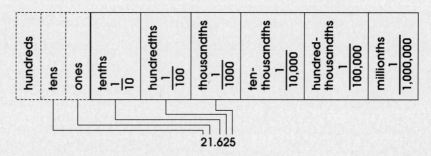

21.625

This number means:

2 tens + 1 ones + 6 tenths + 2 hundredths + 5 thousandths

To read the number, read the value of each section, using the decimal point as the divider.

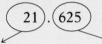

Read: twenty-one and six hundred twenty-five thousandths

To compare decimals, write the numbers with the same number of places. (You can always add zeros at the end of a decimal without changing its value.)

Compare: 0.4   and   0.01
                ↓   ↓
               40   01
            greater

Compare: 0.035   and   0.0912
                ↓   ↓
              0350   0912
                           greater

Compare: 1.06   and   1.007
                ↓   ↓
              1060   1007
            greater

ADDING DECIMALS

Adding decimals is like adding whole numbers. The only trick is to line up the decimal points.

$$6.18 + 3.24 =$$

Step 1: Write the numbers with the decimal points lined up.

$$\begin{array}{r} \downarrow \\ 6.18 \\ +3.24 \\ \hline \end{array}$$

Step 2: Add. Start with hundredths.

8 + 4 = 12

You can't write 12 in the hundredths place. Regroup. Write the 2. Write the 1 over the tenths column.

$$\begin{array}{r} 1 \\ 6.18 \\ +3.24 \\ \hline 2 \end{array}$$

Step 3: Add the tenths.

1 + 1 + 2 = 4

Add the ones.

6 + 3 = 9

Write the decimal point in the answer.

$$\begin{array}{r} 1 \\ 6.18 \\ +3.24 \\ \hline 9\downarrow42 \\ 9.42 \end{array}$$

In adding decimals, be sure to line up the decimal points. Look at how the decimal points are lined up in each of these examples.

$$
\begin{array}{r} 2.01 \\ +13.9 \\ \hline 15.91 \end{array}
\qquad
\begin{array}{r} 697.32 \\ +\ 1.048 \\ \hline 698.368 \end{array}
\qquad
\begin{array}{r} 7.003 \\ +0.12 \\ \hline 7.123 \end{array}
\qquad
\begin{array}{r} 0.02 \\ +11.796 \\ \hline 11.816 \end{array}
$$

SUBTRACTING DECIMALS

In subtracting decimals, be sure to line up the decimal points.

$$7.141 - 1.028 =$$

Step 1: Write the numbers with the decimal points lined up.

$$\begin{array}{r} 7.141 \\ -1.028 \end{array}$$

Step 2: Subtract. Start with thousandths. You can't subtract 8 from 1. Regroup. Then subtract.

$$11 - 8 = 3$$

$$\begin{array}{r} {\scriptstyle 3\ 11} \\ 7.1\cancel{4}\cancel{1} \\ -1.028 \\ \hline 3 \end{array}$$

Step 3: Subtract the hundredths.
$$3 - 2 = 1$$
Subtract the tenths.
$$1 - 0 = 1$$
Subtract the ones.
$$7 - 1 = 6$$
Write the decimal point in the answer.

$$\begin{array}{r} {\scriptstyle 3\ 11} \\ 7.1\cancel{4}\cancel{1} \\ -1.028 \\ \hline 6\ \ 113 \\ 6.113 \end{array}$$

In subtracting decimals, be sure to line up the decimal points. Look at how the decimal points are lined up in each of these examples.

$$
\begin{array}{r} 17.09 \\ -\ 3.1 \\ \hline 13.99 \end{array}
\qquad
\begin{array}{r} 241.0083 \\ -\ 0.32 \\ \hline 240.6883 \end{array}
\qquad
\begin{array}{r} 9.900 \\ -3.006 \\ \hline 6.894 \end{array}
$$

MULTIPLYING DECIMALS

When you multiply decimals, first multiply just as with whole numbers. Then place the decimal point.

Add the number of decimal places in each of the factors. The number of decimal places in the answer should be equal to the total number of decimal places.

Example:

$$
\begin{array}{r} \downarrow\downarrow\downarrow \\ 2.146 \\ \downarrow \\ \times\ \ 1.2 \\ \hline 2.5752 \\ \uparrow\uparrow\uparrow\uparrow \end{array}
\qquad
\begin{array}{l} 3 \text{ decimal places} \\ + \\ \underline{1} \text{ decimal places} \\ 4 \text{ decimal places} \end{array}
$$

Here are the steps to follow.

Step 1: Multiply from the right.

7 × 4 = 28

Regroup. Write the 2 in the next column.

Multiply the 1 times the 4 and add the 2 that you regrouped.

Multiply 3 × 4.

3 × 4 = 12

$$\begin{array}{r} 2 \\ 3.17 \\ \times\,0.24 \\ \hline 1268 \end{array}$$

Step 2: Multiply by the 2 tenths. Write the answer. Add.

$$\begin{array}{r} 3.17 \\ \times\,0.24 \\ \hline 1268 \\ 6340 \\ \hline 7608 \end{array}$$

Step 3: Add the decimal places in each factor.

3.17 ⟶ 2 decimal places

0.24 ⟶ 2 decimal places

4 in all

Place the decimal point with 4 decimal places in the answer.

$$\begin{array}{r} 3.17 \\ \times\,0.24 \\ \hline 1268 \\ 6340 \\ \hline 0.7608 \end{array}$$

Here's an example in which you have to write a zero after the decimal point in order to have enough decimal places in the answer.

Step 1: Multiply as usual. Add the products.

$$\begin{array}{r} 2.7 \\ \times\,0.03 \\ \hline 81 \\ 00 \\ \hline 81 \end{array}$$

Step 2: Count the total number of decimal places. There are 3 in all. Write a zero to show 3 decimal places. Write the decimal point.

$$\begin{array}{r} 2.7 \\ \times\,0.03 \\ \hline 81 \\ 00 \\ \hline .081 \\ \downarrow \\ 0.081 \end{array}$$

(Write a zero in the ones place.)

DIVIDING DECIMALS

Placing the decimal point is the only thing that makes dividing decimals different from dividing whole numbers.

Follow these steps:

Step 1: Write the decimal point directly above the decimal point in the number you are dividing.

$$23\overline{)71.76}$$

Step 2: Divide as usual.

$$\begin{array}{r} 3.12 \\ 23\overline{)71.76} \\ \underline{69} \\ 2\ 7 \\ \underline{2\ 3} \\ 46 \\ \underline{46} \\ 0 \end{array}$$

Example: $0.072 \div 9$

Step 1: Place the decimal point. Write zeros above the zeros.

$$\begin{array}{r} 0.0 \\ 9\overline{)0.072} \end{array}$$

Step 2: You can't divide 7 by 9. Write a zero in the answer. Then divide the thousandths.

$$\begin{array}{r} 0.008 \\ 9\overline{)0.072} \\ \underline{72} \\ 0 \end{array}$$

To divide *by* a decimal, follow these steps:

$$1.176 \div 0.21 =$$

Step 1: Move the decimal point in the number you are dividing by. Move the decimal point in the number you are dividing the same number of places.

$$0.21\overline{)1.176}$$

Step 2: Write the decimal point.

$$21\overline{)117.6}$$

Step 3: Divide as usual.

$$\begin{array}{r} 5.6 \\ 21\overline{)117.6} \\ \underline{105} \\ 12\ 6 \\ \underline{12\ 6} \\ 0 \end{array}$$

ROUNDING DECIMALS

To round a decimal number, look at the digit in the place to the right of the place you want to round to.

Round to tenths:

Look at hundredths.
↓
3.16

Round to hundredths:

Look at thousandths.
↓
3.245

If the digit is 5 or greater, round up.
If the digit is less than 5, leave the same.

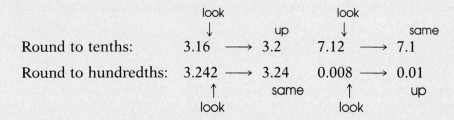

ROUNDING DECIMAL ANSWERS
When you divide decimals, don't show a remainder.

Example: Divide 5.99 by 43. Show your answer in hundredths.

Step 1:   Write the decimal point in the
answer. Divide as usual.

$$
\begin{array}{r}
0.13 \\
43\overline{)5.99} \\
4\,3 \\
\hline
1\,69 \\
1\,29 \\
\hline
40
\end{array}
$$

Step 2:   The answer should be
rounded to hundredths.
Divide to one place beyond
hundredths. There is no digit
in the thousandths place. Add
a zero. Divide thousandths.

$$
\begin{array}{r}
0.139 \\
43\overline{)5.990} \\
4\,3 \\
\hline
1\,69 \\
1\,29 \\
\hline
400 \\
387 \\
\hline
13
\end{array}
$$

Step 3:   Round the answer to hundredths.

look
↓

$$
\begin{array}{r}
\underline{0.139} \quad 0.14 \\
43\overline{)5.990}
\end{array}
$$

When you divide whole numbers, you can show the remainder as a decimal.

Example: Divide 200 by 6. Carry the division to tenths.

Step 1:   Divide as usual. When you get
a remainder, write a decimal
point. Add zeros in the tenths
and hundredths places. Con-
tinue to divide.

$$
\begin{array}{r}
33.33 \\
6\overline{)200.00} \\
18 \\
\hline
20 \\
18 \\
\hline
2\,0 \\
1\,8 \\
\hline
20 \\
18 \\
\hline
2
\end{array}
$$

Step 2:   Round your answer to tenths.

look
↙

$$
\begin{array}{r}
\underline{33.33} \quad 33.3 \\
6\overline{)200.00}
\end{array}
$$

❏ PERCENTS, RATIOS, AND PROPORTIONS

RATIOS

A comparison of two quantities is called a **ratio.** A fraction is a ratio.
In this diagram, there are 4 stars and 8 dots. The ratio of stars to dots
is 4 to 8.

The ratio can be written:

$$4 \text{ to } 8 \qquad 4:8 \qquad \text{or} \qquad \frac{4}{8}$$

The two numbers in a ratio are the **terms** of the ratio.

As with a fraction, a ratio can be written in lowest terms. The fraction ⁴⁄₈ can be written as ½. Write the ratio as:

$$1 \text{ to } 2 \qquad 1:2 \qquad \text{or} \qquad \frac{1}{2}$$

Rectangle A has sides of 1 foot and 3 feet. The ratio of the short side to the longer side is 1:3. Rectangle B has sides of 2 feet and 6 feet. The ratio of the short side to the long side is ²⁄₆ or ⅓. The ratios are **equal.**

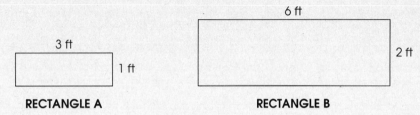

RECTANGLE A   RECTANGLE B

To find an equal ratio, multiply each term by the same number. These ratios are equal:

$$\frac{1 \ (\times \ 2)}{2 \ (\times \ 2)} = \frac{2}{4} \qquad \frac{1 \ (\times \ 5)}{2 \ (\times \ 5)} = \frac{5}{10} \qquad \frac{1 \ (\times \ 8)}{2 \ (\times \ 8)} = \frac{8}{16}$$

If the sides of rectangle A are kept in the same ratio, and the short side is enlarged to 6 feet, how long will the longer side be? Set up equal ratios and find the missing number.

$$\frac{1}{3} = \frac{6}{?} \qquad \frac{1 \ (\times \ 6)}{3 \ (\times \ 6)} = \frac{6}{18}$$

(Think: What number was 1 multiplied by?)

(Multiply 3 by that number.)

The longer side will be 18 feet long.

PROPORTIONS

A **proportion** says that two ratios are equal.

Examples of proportions: $\dfrac{2}{3} = \dfrac{4}{6} \qquad \dfrac{2}{10} = \dfrac{1}{5} \qquad \dfrac{1}{2} = \dfrac{4}{8}$

In a proportion, the cross products are equal.

$\dfrac{3}{6} \diagdown\diagup \dfrac{1}{2}$ $6 \times 1 = 6$

$3 \times 2 = 6$

6 = 6 The ratios are a proportion because the cross products are equal.

To find cross products, multiply the upper term of one ratio times the lower term of the other. Multiply the lower term times the upper term.

Use cross products to solve problems involving proportions.

Problem: It takes 2 eggs to make pancakes for 3 people. How many eggs will it take to make pancakes for 12 people?

Step 1:   Write a proportion. Use *n* for the unknown number. Be sure the ratios name the items in the same order.

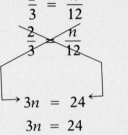

$$\frac{\text{egg}}{\text{people}} = \frac{\text{egg}}{\text{people}}$$

$$\frac{2}{3} = \frac{n}{12}$$

Step 2:   Multiply to get the cross products. Set them equal to each other.

$$3n = 24$$

Step 3:   To solve the equation, divide each side by the same number. The missing number is 8 eggs.

$$3n = 24$$
$$\div 3 \quad \div 3$$
$$n = 8$$

You can solve the problem with equal ratios instead. Follow these steps:

Step 1:   Write the proportion.

$$\frac{2}{3} = \frac{n}{12}$$

Step 2:   Decide what number the lower term was multiplied by. Multiply the upper term by the same number.

$$\frac{2\,(\times 4)}{3\,(\times 4)} = \frac{8}{12}$$
$$n = 8$$

PERCENTS

A **percent** is a ratio where the second term is 100. Percents are written with the percent sign (%). These three ratios are equal. The third is a percent.

$$59:100 \qquad 59 \text{ to } 100 \qquad 59\% \qquad \text{(Say 59 percent.)}$$

A percent can be written as a fraction with 100 in the denominator. It can also be written as a decimal.

$$50\% = \frac{50}{100} = 0.50 \qquad \text{(The number 0.50 means 50 hundredths or } {}^{50}/_{100}.)$$

To change a percent to a decimal, move the decimal point two places to the left, and drop the percent sign.

$$37\% = 0.37 \qquad 115\% = 1.15 \qquad 3567\% = 35.67$$

(These are the steps: $37\% = {}^{37}/_{100} = 0.37$.)

You can also change any decimal to a percent.

To change a decimal to a percent, move the decimal point two places to the right and add the percent sign.

$$0.89 = 89\% \qquad 1.73 = 173\% \qquad 0.01 = 1\%$$

(These are the steps: $0.89 = {}^{89}/_{100} = 89\%$.)

You can write any ratio as a percent. Use cross products.

One out of ten people is left-handed. What percent is that? (Remember, a percent has 100 in the denominator.)

$$\frac{1}{10} \diagup\!\!\!\!\diagdown = \diagup\!\!\!\!\diagdown \frac{n}{100} \qquad 10n = 100$$
$$\div 10 \quad \div 10$$
$$n = 10 \qquad 10 \text{ percent}$$

This formula can be used to solve percentage problems.

percent × base = amount

Examples of the formula:

$25\% \times 8 = 2$ $\qquad\qquad$ $30\% \times 100 = 30$

(25 percent of 8 is 2.) $\qquad$ (30 percent of 100 is 30.)

$15\% \times 60 = 9$ $\qquad\qquad$ $75\% \times 200 = 150$

(15 percent of 60 is 9.) $\qquad$ (75 percent of 200 is 150.)

If you know two of the numbers in the formula, you can find the third.

FINDING THE AMOUNT

To find 30 percent of 500, follow these steps:

Step 1: Write the numbers you know and the unknown number in the percent formula. (Think 30 percent of 500 is ____.)

percent × base = amount
$30\% \times 500 = n$

Step 2: Solve the formula. Multiply. To multiply by a percent, change it to a decimal first.

$30\% \times 500 = n$
$\downarrow$
$.30 \times 500 = n$

$$\begin{array}{r} 500 \\ \times\ .30 \\ \hline 150.00 \end{array}$$

FINDING THE BASE

To solve a problem such as "12 percent of what number is 3.6?" follow these steps.

Step 1: Write the numbers you know and the unknown number in the percent formula. This time, you know the amount (3.6) and you know the percent (12).

(Think 12 percent of ____ is 3.6.)

percent × base = amount
$12\% \times n = 3.6$

Step 2: To solve the equation, divide each side by the same number—12 percent. First convert 12 percent to a decimal.

$12\% \times n = 3.6$
$\downarrow$
$.12 \times n = 3.6$

Step 3: Divide.

$$.12 \times n =\ \ 3.6$$
$$\div .12 \quad \div .12$$
$$n = 30$$

FINDING THE PERCENT

To solve a problem such as "What percent is 16 of 64?" follow these steps:

Step 1: Write the numbers you know and the unknown number in the percent formula.

(Think _____ percent of 64 is 16.)

$$\text{percent} \times \text{base} = \text{amount}$$
$$n \times 64 = 16$$

Step 2: Divide each side by the same number. Change the decimal to a percent.

$$n \times 64 = 16$$
$$\div 64 \quad \div 64$$
$$n = .25$$
$$25\%$$

$$\begin{array}{r} .25 \\ 64\overline{)16.00} \\ 12\ 8 \\ \hline 3\ 20 \\ 3\ 20 \end{array}$$

FINDING PERCENT OF INCREASE OR DECREASE

Many percent problems involve an increase or decrease. For example, a quantity has increased from 40 to 60. What percent of increase was that?

Follow these steps to find a percent of increase:

Step 1: First find the amount of increase.

$$60 - 40 = 20$$

The amount of increase is 20.

$$\begin{array}{r} 60 \\ -40 \\ \hline 20 \end{array}$$

Step 2: Write the numbers you know in the formula. Use the original number as the base. Divide to solve.

$$\text{percent} \times \text{base} = \text{amount}$$
$$n \times 40 = 20$$
$$\div 40 \quad \div 40$$
$$n = .5$$

Step 3: Convert the decimal to a percent. The answer is that 40 to 60 is a 50-percent increase.

$$.50 = 50\%$$

Suppose a quantity has decreased from 60 to 40. What is the percent of decrease from 60 to 40?

Follow these steps to find a percent of decrease:

Step 1: First determine how much the quantity decreased.

$$60 - 40 = 20$$

$$\begin{array}{r} 60 \\ -40 \\ \hline 20 \end{array}$$

Step 2: Write the numbers you know in the formula. Use the original number as the base. Divide out to thousandths. Round to hundredths.

$$\text{percent} \times \text{base} = \text{amount}$$
$$n \times 60 = 20$$
$$\div 60 \quad \div 60$$
$$n = .333$$

Step 3: Convert the decimal to a percent. The answer is that 60 to 40 is a 33-percent decrease.

$$.33 = 33\%$$

DISCOUNT AND MARKUP

Many problems with percents have to do with discounts and markups.

> Example: An item is marked "25% off." The original price is $100. What is the new price?

Follow these steps:

Step 1: First find the discount. The discount is the amount that the price will be reduced. It is 25 percent of the original price.

(25 percent of $100 is ____.)

$$\text{percent} \times \text{base} = \text{amount}$$
$$25\% \times 100 = n$$
$$.25 \times 100 = n$$
$$25 = n$$

Step 2: Subtract the discount from the original price.

$$\begin{array}{r} \$100 \\ -\ \ 25 \\ \hline \$\ 75 \end{array}$$

> Example: An item costs $25. The store selling it will mark up the price by 15 percent. What will the selling price be?

Follow these steps:

Step 1: The markup is the amount the price will be raised. It is 15 percent of the original price.

(15 percent of $25 is ____.)

$$\text{percent} \times \text{base} = \text{amount}$$
$$15\% \times 25 = n$$
$$.15 \times 25 = n$$
$$3.75 = n$$

Step 2: Add the markup to the original price.

$$\begin{array}{r} \$25.00 \\ +\ \$\ 3.75 \\ \hline \$28.75 \end{array}$$

❏ DESCRIPTIVE STATISTICS

Descriptive statistics are ways to describe data. It's often useful to describe data in two ways. One is by **measures of central tendency** (how data group around certain numbers). Other useful descriptions are **measures of variability** (how data are spread or scattered).

❏ Example of a statement that is a measure of central tendency: "The average number of students per class is 27."

❏ Example of a statement that is a measure of variability: "The largest class has 45, and the smallest class has 16."

MEASURES OF CENTRAL TENDENCY

Mean   The sum of all the data divided by the number of data. (Sometimes called the "average.")

Median   The middle number when the data are arranged in numerical order.

Mode   The number that occurs most often.

Here is how to find the mean, the median, and the mode for this set of data.

$$84 \quad 86 \quad 84 \quad 87 \quad 92 \quad 95 \quad 67$$

To find the mean:

Step 1: Add the numbers.
$$84 + 86 + 84 + 87 + 92 + 95 + 67 = 595$$

Step 2: Count the data. There are 7 in all. Divide by the number of data.

$$\begin{array}{r} 85 \\ 7\overline{)595} \\ \underline{56} \\ 35 \end{array}$$

mean = 85

To find the median:

Step 1: Arrange the numbers in order from least to greatest.

$$67 \quad 84 \quad 84 \quad 86 \quad 87 \quad 92 \quad 95$$

Step 2: Find the middle number. The number of data is 7, so the fourth number is the middle one.

$$67 \quad 84 \quad 84 \quad 86 \quad 87 \quad 92 \quad 95$$
$$\uparrow$$
median = 86

To find the median when you have an even number of data:

Step 1: Arrange the numbers in order and look for the middle number. There are 8 numbers, so no score falls right in the middle. Pick the 2 numbers just above and below the middle.

$$9 \; 10 \; 10 \; 12 \; 14 \; 17 \; 17 \; 19$$
$$12 \; 14$$

Step 2: Find the mean (average) of these 2 numbers. Add them. Divide by 2.

$$12 + 14 = 26$$
$$26 \div 2 = 13$$
median = 13

To find the mode:

Step 1: It usually helps to arrange the data in order. Then count to see how many of each number you have.

$$67 \quad 84 \quad 84 \quad 86 \quad 87 \quad 92 \quad 95$$
$$1 \qquad 2 \qquad\quad 1 \quad\; 1 \quad\; 1 \quad\; 1$$

Step 2: The number that occurs most often is the mode.

mode = 84

MEASURES OF VARIABILITY

The most common measure of variability is range.

Range The difference between the highest and the lowest numbers in a group of data.

Step 1: Again, it helps to arrange the data in order. Find the lowest and the highest number.

$$67 \quad 84 \quad 84 \quad 86 \quad 87 \quad 92 \quad 95$$

Step 2: Subtract the lowest number from the highest number. The difference is the range.

$$95 - 67 = 28$$
$$\text{range} = 28$$

❏ GRAPHS AND TABLES

Graphs and tables organize and present information. Things that can be counted and that can be divided in some way can be shown in a graph or table.

A table contains numbers arranged in rows and columns. The rows and columns have labels. This table shows how much steel a factory produced over a 5-month period.

Example of a Completed Table:

TABLE — FACTORY STEEL PRODUCTION

MONTH	TONS OF STEEL
JAN	23
FEB	40
MAR	26
APR	30
MAY	40

A graph shows information visually. This graph shows the same information.

Example of a Completed Graph:

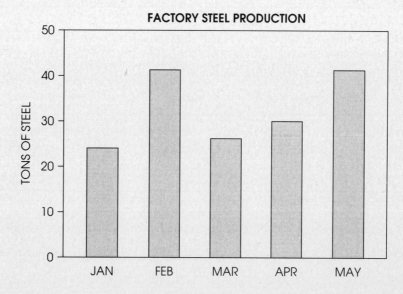

The graph lets you make comparisons easily. The tallest bar shows the greatest production. The smallest bar shows the least production. The graph also lets you see trends. There is a big drop in month 3 and then a return in month 5 to the highest level of production.

The table lets you see the numbers clearly. Comparisons and trends are not as clear. Often you make a table before you make a graph to help organize the numbers first.

Different kinds of graphs are shown below.

Examples of Different Kinds of Graphs:

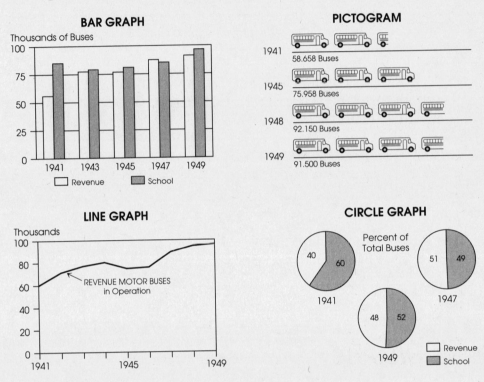

BAR GRAPHS

Bar graphs are used to show comparison. A bar graph has a **vertical axis** and a **horizontal axis.** It has bars showing data.

Example of a Completed Bar Graph:

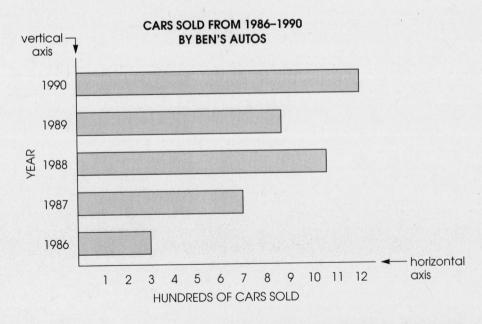

This graph shows the number of cars sold by a dealer from 1986 to 1990.

To read any bar graph, follow these steps:

❏ Read the title of the graph to see what the graph is about. This graph shows car sales for a dealer.

❏ Read the label on each axis. The vertical axis shows years from 1986 through 1990. The horizontal axis shows cars sold. Notice that each number of the horizontal axis stands for that many hundreds of cars.

❏ To read any data on the graph, find the label and the bar that goes with it. Then trace a line from the end of the bar to the other axis.

❏ To see how many cars the dealer sold in any year, find the label for the year. The bar stands for the cars sold that year. For 1986, the end of the bar is at 3, which stands for 300 cars. For 1989, the end of the bar is three-fourths of the way from 8 to 9 or about 8.75. It stands for 875 cars.

❏ The longest bar is the greatest number. The shortest bar is the least number.

The year with fewest sales was 1986.
The year with most sales was 1990.

LINE GRAPHS

Line graphs are used to show trends over time. A line graph also has two axes. It shows data with a line.

Example of a Completed Line Graph:

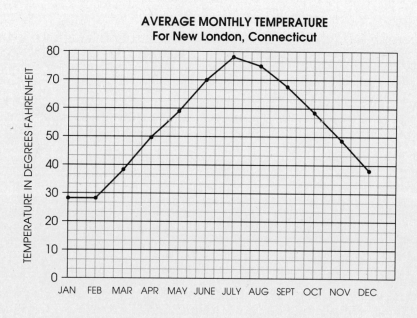

This graph shows temperatures.

To read any line graph, follow these steps:

❑ Read the title of the graph to see what the graph is about. This graph shows the average monthly temperature for New London, Connecticut.

❑ Read the label on each axis. The vertical axis shows temperature in degrees from 0 to 80. Notice that only every tenth degree is labeled. The horizontal axis shows months.

❑ To read the data, find the correct dot. Trace from the dot to the axis to read the data. To find the temperature in May, find the label "May." Locate the dot above May. Trace it to the vertical axis. It is at just below 60, or about 58° F. The dot for April is at 50° F.

❑ The line shows trends up and down. The labels on the graph tell when the trend started up or down. The average monthly temperature rises from February through July and then falls to the end of the year.

CIRCLE GRAPHS

Circle graphs are used to compare parts of a whole. A circle graph is a circle, with data shown as parts (wedges) of the circle.

Example of a Completed Circle Graph:

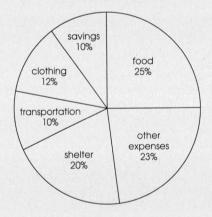

HOW A TYPICAL AMERICAN FAMILY SPENDS ITS INCOME

This graph shows how income is spent.

To read any circle graph, follow these steps:

❑ Read the title of the graph to see what the graph is about. This graph shows how a typical family spends its income.

❑ Read the label on each section of the circle. It identifies the section and gives a percentage. The graph shows the percentage spent on clothing, savings, food, and so on.

❑ A circle graph shows how a total quantity is divided. The sections of the graph always total 100 percent. To read the data, find the correct section. Read the percentage in the section. To find the percent of total income that a family spends on savings, find the savings section. The percentage shown is 10 percent.

❏ The circle graph shows comparisons between sizes of items. The sections for transportation and savings are the same. Each is 10 percent. The largest section is food, at 25 percent. Even if the graph labels don't include percentages, you can compare the sections. You can estimate how large each is.

❏ GEOMETRY

There are many different kinds of geometric figures. Here are some common ones:

parallelogram A figure with four sides; the opposite sides are equal and parallel to each other.

rectangle A special kind of parallelogram, with four right angles. Opposite sides are equal and parallel.

square A special kind of rectangle, with four equal sides and four right angles.

triangle A figure with three sides.

right triangle A triangle with one right angle.

PERIMETER

The **perimeter** is the total distance around a figure. It is the sum of the sides.

To find the perimeter of a triangle, add the lengths of the sides.

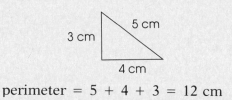

perimeter = 5 + 4 + 3 = 12 cm

To find the perimeter of a parallelogram or a rectangle, add the lengths of the sides.

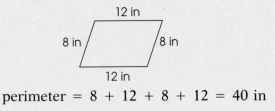

12 in

$$\text{perimeter} = 8 + 12 + 8 + 12 = 40 \text{ in}$$

You can find the perimeter of a parallelogram or rectangle if you know the length of one shorter side and one longer side.

3 ft

6 ft

The opposite sides of a parallelogram or rectangle are equal. So you know the other two sides are also 6 feet and 3 feet.

$$\text{perimeter} = 6 + 3 + 6 + 3 = 18 \text{ ft}$$

Since the sides of a square are equal, you can find the perimeter by multiplying 4 times the length of any side.

15 ft

$$\text{perimeter} = 4 \times 15 = 60 \text{ ft}$$

AREA

Area is the number of square units needed to cover a figure. A square unit is a square inch (sq in), square centimeter (sq cm), square yard (sq yd), square mile (sq mi), and so on.

A square centimeter is 1 centimeter on a side. A square inch is one inch on a side. A square yard is 1 yard on a side.

SQUARE CENTIMETER

This rectangle is covered by 8 square feet. Its area is 8 square feet.

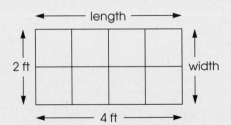

length

2 ft

width

4 ft

To find the area of a rectangle, use this formula: $A = l \times w$

A stands for *area*; l stands for *length*; w stands for *width*.

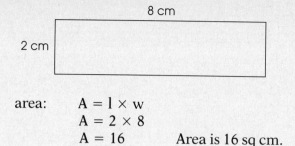

8 cm

2 cm

area: A = l × w
A = 2 × 8
A = 16   Area is 16 sq cm.

To find the area of a square, again multiply the length times the width. Since the sides of a square are equal, you multiply the length of one side times itself.

4 in

area: A = s × s
A = 4 × 4
A = 16   Area is 16 sq in.

This diagram shows a rectangle cut in half. Each half is a triangle. The area of each half of the rectangle is equal. Looking at this diagram helps you to understand how to find the area of a triangle.

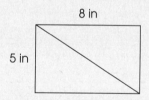

8 in

5 in

The area of the rectangle is: A = l × w
A = 5 × 8 = 40 sq in

The area of each triangle is half of this, or 20 square inches.

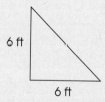

6 ft

6 ft

One way to write the formula for the area of a triangle is:

$$A = \frac{1}{2} \times (l \times w) \leftarrow \text{half the area of a rectangle}$$

$$A = \frac{1}{2} \times 6 \times 6 = \frac{1}{2} \times 36 = 18 \text{ sq ft}$$

This formula works because the triangle is a right triangle, and it is half of a rectangle. What about other triangles? The usual formula uses the base of the triangle and its height.

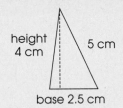

The shortest side of a triangle is the base. The shortest distance from the base to the highest point of the triangle is the height.

$$A = \frac{1}{2}(b \times h)$$

$$A = \frac{1}{2}(2.5 \times 4)$$

$$A = \frac{1}{2}(10) = 5 \qquad \text{Area is 5 sq cm.}$$

The formula for the area of a parallelogram is:

$$A = b \times h$$

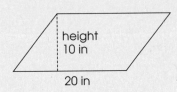

$$A = b \times h$$
$$A = 20 \times 10 = 200 \qquad \text{Area is 200 sq in.}$$

CIRCLES

A circle is a closed curve with all points equally distant from the center. The distance from the center of the circle to the rim of the circle is the **radius** of the circle.

The distance around a circle is the **circumference.** The formula for circumference uses the Greek symbol π, which is spelled *pi*. It is pronounced "pie." Pi always has the same value.

$$\pi = \text{approximately 3.14 or } \frac{22}{7}$$

The formula for circumference of a circle is:

$$C = 2 \times \pi r$$

C is the circumference; *r* is the radius.

The circumference of the circle is: $C = 2 \times \pi \times 5$
$C = 2 \times 3.14 \times 5$
$C = 31.4$ in

The formula for area of a circle is: $A = \pi \times r \times r$

The area of the circle above is: $A = \pi \times r \times r$
$A = 3.14 \times 5 \times 5$
$A = 78.5$ sq in

VOLUME

Squares, rectangles, triangles, and circles are flat figures. Figures that have three dimensions are **solid figures.** Each surface of a solid figure is a **face.**

Prism   A solid figure with parallel bases. The other faces are parallelograms.

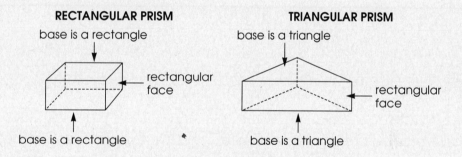

Cube   A solid figure with square bases and faces.

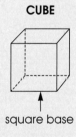

Cylinder A solid figure with two bases that are circles.

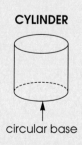

The volume of a solid figure is the number of cubic units it holds. A cubic unit is a cubic inch (cu in), cubic centimeter (cu cm), and so on.

CUBIC CENTIMETER

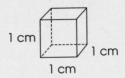

1 cm
1 cm
1 cm

To find the volume of a prism, use this formula:

$$V = l \times w \times h$$

V is volume; l is length; w is width; h is height.

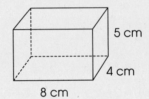

5 cm
4 cm
8 cm

$V = l \times w \times h$
$V = 8 \times 4 \times 5 = 160$ Volume is 160 cu cm.

To find the volume of a cube, use the same formula. The length, width, and height of a cube are all the same.

6 in

$V = l \times w \times h$
$V = 6 \times 6 \times 6 = 216$ Volume is 216 cu in.

To find the volume of a cylinder, first find the area of the base and then multiply times the height. The formula is:

$$V = (\pi \times r \times r) \times h$$

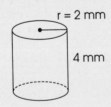

r = 2 mm
4 mm

$V = (\pi \times r \times r) \times h$
$V = (3.14 \times 2 \times 2) \times 4$
$V = 12.56 \times 4 = 50.24$ Volume is 50.24 cu mm.

Mathematics Review Skills Table

First, use the Answer Key on page 145 to correct your test. Then, fill in an X in the second column next to each answer that is wrong. Look at the page numbers in the third column next to each X. The third column tells you which pages you should study to help build your skills.

Question	Put an **X** for a wrong answer	Study these pages
1		Pages 105 – 115
2		
3		
4		
5		Pages 115 – 121
6		
7		
8		
9		
10		Pages 121 – 126
11		
12		
13		Pages 126 – 131
14		
15		
16		
17		Pages 131 – 133
18		Pages 137 – 142
19		
20		

Answer Key

MATH PRETEST

1. 390 cars
2. 375 tires
3. 120 hours
4. 22 miles per gallon
5. 42⅛ in.
6. 7¾ in.
7. 8½ gal.
8. 20 transmissions
9. Battery: $52   Filter: $6.15
   Total cost: $58.15
10. $4.74

11. $44.95
12. 0.10
13. 12 cars
14. 25%
15. 16%
16. $48 a dozen
17. mean = $9.23
    median = $8.85
18. 36 feet
19. 121 square feet
20. 282.6 cubic feet

Answer Key

JOB SITUATION 1

TASK 1

1. **To Do:** Find Angela Upton's gross pay. Write the total hours and gross pay in the payroll register.

   Steps:
   1. Find the total number of hours worked.
   2. Find and record the gross pay.

   Information: Hourly Wage: $7.75 per hour.

   Hours Worked:
Monday	8 hours
Tuesday	8 hours
Wednesday	8 hours
Thursday	8 hours
Friday	8 hours

   System of Measurement: Time and money

   Compute:
   $$8 + 8 + 8 + 8 + 8 = 40 \text{ hours}$$
   $$40 \times \$7.75 = \$310$$

   Communicate: See the completed payroll register on page 146.

2. **To Do:** Find Patrick O'Leary's gross pay. Write the total hours and gross pay in the payroll register.

   Steps:
   1. Find the total number of hours worked.
   2. Find and record the gross pay.

   Information: Hourly Wage: $10.70 per hour.

   Hours Worked:
Monday	9 hours
Tuesday	7 hours
Wednesday	8 hours
Thursday	9 hours
Friday	6 hours

   System of Measurement: Time and money

   Compute:
   $$9 + 7 + 8 + 9 + 6 = 39 \text{ hours}$$
   $$39 \times \$10.70 = \$417.30$$

   Communicate: See the completed payroll register on page 146.

3. **To Do:** Find Jamal Cheston's gross pay. Write the total hours and gross pay in the payroll register.

   Steps:
   1. Find the total number of hours worked.
   2. Find and record the gross pay.

   Information: Hourly Wage: $11.00 per hour.

   Hours Worked:
Monday	5 hours
Tuesday	4 hours
Wednesday	5 hours
Thursday	6 hours
Friday	5 hours

   System of Measurement: Time and money

   Compute:
   $$5 + 4 + 5 + 6 + 5 = 25 \text{ hours}$$
   $$25 \times \$11.00 = \$275.00$$

   Communicate:
   See the completed payroll register below.

4. **To Do:** Find Barbara Palmer's gross pay. Write the total hours and gross pay in the payroll register.

   Steps:
   1. Find the total number of hours worked.
   2. Find and record the gross pay.

   Information: Hourly Wage: $8.25 per hour.

   Hours Worked:
Monday	8 hours
Tuesday	9 hours
Wednesday	9 hours
Thursday	9 hours
Friday	8 hours

   System of Measurement: Time and money

   Compute:
   $$8 + 9 + 9 + 9 + 8 = 43 \text{ hours}$$
   $$43 \times \$8.25 = \$354.75$$

   Communicate:
   See the completed payroll register below.

5. **To Do:** Find Sam Li's gross pay. Write the total hours and gross pay in the payroll register.

   Steps:
   1. Find the total number of hours worked.
   2. Find and record the gross pay.

   Information: Hourly Wage: $9.20 per hour.

   Hours Worked:
Monday	7 hours
Tuesday	8 hours
Wednesday	9 hours
Thursday	7½ hours
Friday	8 hours

   System of Measurement: Time and money

   Compute:
   $$7 + 8 + 9 + 7\tfrac{1}{2} + 8 = 39\tfrac{1}{2} \text{ hours}$$
   $$39\tfrac{1}{2} \times \$9.20 = 39.5 \times 9.20 = \$363.40$$

   Communicate:
   See the completed payroll register below.

PAYROLL REGISTER   Week Ending 1/10/9X

| EMPLOYEE INFORMATION | | | | | GROSS EARNINGS | | DEDUCTIONS | | |
NAME	MAR STAT	ALLOW	TOTAL HOURS	REG RATE	GROSS PAY	FICA	FWT	NET PAY
Peggy Laskowitz	M	4	40	8.50	340.00			
Angela Upton	S	1	40	7.75	310.00			
Patrick O'Leary	S	1	39	10.70	417.30			
Jamal Cheston	M	5	25	11.00	275.00			
Barbara Palmer	S	2	43	8.25	354.75			
Sam Li	M	3	39½	9.20	363.40			

TASK 2

6. **To Do:** Find the FICA tax for Angela Upton.

   Steps: Multiply gross pay by 7.65%

   Information: Gross pay is $310.00.

   System of Measurement: Money

   Compute:
   $$\$310.00 \times 0.0765 = \$23.715 = \$23.72$$

   Communicate: See the completed payroll register on page 157.

7. **To Do:** Find the FICA tax for Patrick O'Leary.

   Steps: Multiply gross pay by 7.65%

   Information: Gross pay is $417.30.

   System of Measurement: Money

   Compute:
   $417.30 \times 0.0765 = \$31.92$

   Communicate: See the completed payroll register below.

8. **To Do:** Find the FICA tax for Jamal Cheston.

   Steps: Multiply gross pay by 7.65%

   Information: Gross pay is $275.00.

   System of Measurement: Money

   Compute:
   $275.00 \times 0.0765 = \$21.0375 = \$21.04$

   Communicate: See the completed payroll register below.

9. **To Do:** Find the FICA tax for Barbara Palmer.

   Steps: Multiply gross pay by 7.65%

   Information: Gross pay is $354.75.

   System of Measurement: Money

   Compute:
   $354.75 \times 0.0765 = \$27.138 = \$27.14$

   Communicate: See the completed payroll register below.

10. **To Do:** Find the FICA tax for Sam Li.

    Steps: Multiply gross pay by 7.65%

    Information: Gross pay is $363.40.

    System of Measurement: Money

    Compute:
    $363.40 \times 0.0765 = \$27.80$

    Communicate: See the completed payroll register below.

PAYROLL REGISTER   Week Ending 1/10/9X

NAME	MAR STAT	ALLOW	TOTAL HOURS	REG RATE	GROSS PAY	FICA	FWT	NET PAY
Peggy Laskowitz	M	4	40	8.50	340.00	26.01		
Angela Upton	S	1	40	7.75	310.00	23.72		
Patrick O'Leary	S	1	39	10.70	417.30	31.92		
Jamal Cheston	M	5	25	11.00	275.00	21.04		
Barbara Palmer	S	2	43	8.25	354.75	27.14		
Sam Li	M	3	39½	9.20	363.40	27.80		

EMPLOYEE INFORMATION · GROSS EARNINGS · DEDUCTIONS

TASK 3

11. **To Do:** Figure FWT for Angela Upton and record it in the payroll register.

    Steps:
    1. Find the FWT in the tax tables.
    2. Record the tax.

    Information:
    Gross pay $310.00; single; 1 allowance

    System of Measurement: Money

    Compute:
    From the Single Persons Weekly Payroll Period, 1 allowance = $36.

    Communicate: See the completed payroll register on page 148.

12. **To Do:** Figure FWT for Patrick O'Leary and record it in the payroll register.

    Steps:
    1. Find the FWT in the tax tables.
    2. Record the tax.

    Information:
    Gross pay $417.30; single; 1 allowance

    System of Measurement: Money

    Compute:
    From the Single Persons Weekly Payroll Period, 1 allowance = $52.

    Communicate: See the completed payroll register on page 148.

13. **To Do:** Figure FWT for Jamal Cheston and record it in the payroll register.

   Steps:
   1. Find the FWT in the tax tables.
   2. Record the tax.

   Information:
   Gross pay $275.00; married; 5 allowances

   System of Measurement: Money

   Compute:
   From the Married Persons Weekly Payroll Period, 5 allowances = $0.

   Communicate:
   See the completed payroll register below.

14. **To Do:** Figure FWT for Barbara Palmer and record it in the payroll register.

   Steps:
   1. Find the FWT in the tax tables.
   2. Record the tax.

   Information:
   Gross pay $354.75; single; 2 allowances.

   System of Measurement: Money

   Compute:
   From the Single Persons Weekly Payroll Period, 2 allowances = $37.

   Communicate:
   See the completed payroll register below.

15. **To Do:** Figure FWT for Sam Li and record it in the payroll register.

   Steps:
   1. Find the FWT in the tax tables.
   2. Record the tax.

   Information:
   Gross pay $363.40; married; 3 allowances.

   System of Measurement: Money

   Compute:
   From the Married Persons Weekly Payroll Period, 3 allowances = $26.

   Communicate:
   See the completed payroll register below.

PAYROLL REGISTER   Week Ending 1/10/9X

	EMPLOYEE INFORMATION			GROSS EARNINGS		DEDUCTIONS		
NAME	MAR STAT	ALLOW	TOTAL HOURS	REG RATE	GROSS PAY	FICA	FWT	NET PAY
Peggy Laskowitz	M	4	40	8.50	340.00	26.01	15.00	
Angela Upton	S	1	40	7.75	310.00	23.72	36.00	
Patrick O'Leary	S	1	39	10.70	417.30	31.92	52.00	
Jamal Cheston	M	5	25	11.00	275.00	21.04	0.00	
Barbara Palmer	S	2	43	8.25	354.75	27.14	37.00	
Sam Li	M	3	39½	9.20	363.40	27.80	26.00	

TASK 4

16. **To Do:** Figure take-home pay for Angela Upton and write it in the payroll register.

   Steps:
   1. Find the total deductions.
   2. Subtract the total deductions from the gross pay.

   Information: Gross pay $310.00; FICA = $23.72; FWT = $36

   System of Measurement: Money

   Compute:
   $23.72 + $36 = $59.72
   $310 - $60.72 = $250.28

   Communicate: See the completed payroll register on page 149.

17. **To Do:** Figure take-home pay for Patrick O'Leary and write it in the payroll register.

   Steps:
   1. Find the total deductions.
   2. Subtract the total deductions from the gross pay.

   Information: Gross pay $417.30; FICA = $31.92; FWT = $52.

   System of Measurement: Money

   Compute:
   $31.92 + $52 = $83.92
   $417.30 - $83.92 = $333.38

   Communicate: See the completed payroll register on page 149.

18. **To Do:** Figure take-home pay for Jamal Cheston and write it in the payroll register.

Steps:
1. Find the total deductions.
2. Subtract the total deductions from the gross pay.

Information: Gross pay $275.00; FICA = $21.04; FWT = $0.

System of Measurement: Money

Compute:

$21.04 + $0 = $21.04
$275.00 − $21.04 = $253.96

Communicate:
See the completed payroll register below.

19. **To Do:** Figure take-home pay for Barbara Palmer and write it in the payroll register.

Steps:
1. Find the total deductions.
2. Subtract the total deductions from the gross pay.

Information: Gross pay $354.75; FICA = $27.14; FWT = $37.

System of Measurement: Money

Compute:

$27.14 + $37 = $64.14
$354.75 − $64.14 = $290.61

Communicate:
See the completed payroll register below.

20. **To Do:** Figure take-home pay for Sam Li and write it in the payroll register.

Steps:
1. Find the total deductions.
2. Subtract the total deductions from the gross pay.

Information: Gross pay $363.40; FICA = $27.80; FWT = $26.

System of Measurement: Money

Compute:

$27.80 + $26 = $53.80
$363.40 − $53.80 = $309.60

Communicate:
See the completed payroll register below.

PAYROLL REGISTER   Week Ending 1/10/9X

| EMPLOYEE INFORMATION | | | | GROSS EARNINGS | | | DEDUCTIONS | | |
NAME	MAR STAT	ALLOW	TOTAL HOURS	REG RATE	GROSS PAY	FICA	FWT	NET PAY
Peggy Laskowitz	M	4	40	8.50	340.00	26.01	15.00	298.99
Angela Upton	S	1	40	7.75	310.00	23.72	36.00	250.28
Patrick O'Leary	S	1	39	10.70	417.30	31.92	52.00	333.38
Jamal Cheston	M	5	25	11.00	275.00	21.04	0.00	253.96
Barbara Palmer	S	2	43	8.25	354.75	27.14	37.00	290.61
Sam Li	M	3	39½	9.20	363.40	27.80	26.00	309.60

JOB SITUATION 2

TASK 1

1. **To Do:** Fill in the information on the work order.

Steps:
1. Fill in the customer and car information.
2. Fill in the labor description.
3. List the parts.

Information:
Name: Alan Billings
Address: 100 Plymouth Ave.
     Northtown, IL 50050

Phone: 555-3498
90 Ford Taurus
License 4RD 111
Labor: tune-up
Parts: 6 spark plugs 135A

System of Measurement: Time and money

Compute/Communicate:
The work order looks like this (see page 150).

WORK ORDER

DATE: 1/10/9X				SERVICE/LABOR	PRICE
NAME: Alan Billings				Tune-up	
ADDRESS: 100 Plymouth Ave.					
CITY/ZIP: Northtown, IL 50050					
PHONE: 555-3498					

YEAR	MAKE	MODEL	LICENSE	TOTAL LABOR	
90	Ford	Taurus	4RD 111		

QTY.	PART		PRICE	TOTAL LABOR	
6	Spark plugs 135A			TOTAL PARTS	
				TOTAL SALE	
	TOTAL PARTS			CUSTOMER SIGNATURE: _____	

2. **To Do:** Fill in the information on the work order.

Steps:
1. Fill in the customer and car information.
2. Fill in the labor description.
3. List the parts.

Information:
Name: Alice Anderson
Address: 1 Crescent Blvd.
Northtown, IL 50050

Phone: 555-1111
82 Chevrolet Caprice
License 5TY 123
Labor: oil change
Parts: oil filter 12J, 4 qt. oil

System of Measurement: Time and money

Compute/Communicate:
The work order looks like this.

WORK ORDER

DATE: 1/10/9X				SERVICE/LABOR	PRICE
NAME: Alice Anderson				Oil change	
ADDRESS: 1 Crescent Blvd.					
CITY/ZIP: Northtown, IL 50050					
PHONE: 555-1111					

YEAR	MAKE	MODEL	LICENSE	TOTAL LABOR	
82	Chevy	Caprice	5TY 123		

QTY.	PART		PRICE	TOTAL LABOR	
1	Oil filter 12J			TOTAL PARTS	
4	Qt. oil			TOTAL SALE	
	TOTAL PARTS			CUSTOMER SIGNATURE: _____	

3. **To Do:** Fill in the information on the work order.

Steps:
1. Fill in the customer and car information.
2. Fill in the labor description.
3. List the parts.

Information:
Name: Kathryn Kirkpatrick
Address: 34 Old Mill Lane
Northtown, IL 50050

Phone: 555-3498
87 Honda Civic
License 4GT 222
Labor: replace battery
Part: UB-111

System of Measurement: Time and money

Compute/Communicate:
The work order looks like this (page 151).

WORK ORDER

DATE: 1/10/9X				SERVICE/LABOR	PRICE
NAME: Kathryn Kirkpatrick **ADDRESS:** 34 Old Mill Lane **CITY/ZIP:** Northtown, IL 50050 **PHONE:** 555-3498				Check battery Replace battery	
YEAR	MAKE	MODEL	LICENSE		
87	Honda	Civic	4GT 222	**TOTAL LABOR**	
QTY.	PART		PRICE	**TOTAL LABOR** **TOTAL PARTS**	
1	Battery UB-111			**TOTAL SALE**	
	TOTAL PARTS			**CUSTOMER SIGNATURE:** _____	

4. **To Do:** Fill in the information on the work order.

   Steps:
   1. Fill in the customer and car information.
   2. Fill in the labor description.
   3. List the parts.

   Information:
   Name: Herb Jones
   Address: 111 West Birch
       Northtown, IL 50050

   Phone: 555-7634
   79 Dodge Dart
   License 367 UUY
   Labor: replace water pump
   Part: 328Q

   System of Measurement: Time and money
   Compute/Communicate:
   The work order looks like this.

WORK ORDER

DATE: 1/10/9X				SERVICE/LABOR	PRICE
NAME: Herb Jones **ADDRESS:** 111 West Birch **CITY/ZIP:** Northtown, IL 50050 **PHONE:** 555-7634				Replace water pump	
YEAR	MAKE	MODEL	LICENSE		
79	Dodge	Dart	367 UUY	**TOTAL LABOR**	
QTY.	PART		PRICE	**TOTAL LABOR** **TOTAL PARTS**	
1	Water pump 328Q			**TOTAL SALE**	
	TOTAL PARTS			**CUSTOMER SIGNATURE:** _____	

5. **To Do:** Fill in the information on the work order.

   Steps:
   1. Fill in the customer and car information.
   2. Fill in the labor description.
   3. List the parts.

   Information:
   Name: Sam Avino
   Address: 527 East Elm
       Northtown, IL 50050

   Phone: 555-1287
   88 Buick Regal
   License SAMMY
   Labor: adjust brakes

   System of Measurement: Time and money
   Compute/Communicate:
   The work order looks like this (see page 152).

WORK ORDER

DATE: 1/10/9X				SERVICE/LABOR	PRICE
NAME: Sam Avino				Adjust brakes	
ADDRESS: 527 East Elm					
CITY/ZIP: Northtown, IL 50050					
PHONE: 555-1287					
YEAR	MAKE	MODEL	LICENSE		
88	Buick	Regal	SAMMY	TOTAL LABOR	
QTY.	PART		PRICE	TOTAL LABOR	
				TOTAL PARTS	
				TOTAL SALE	
		TOTAL PARTS		CUSTOMER SIGNATURE: _____	

TASK 2

6. **To Do:** Fill in the costs on the work order.

   Steps:
   1. Find and record the labor cost.
   2. Find the parts cost in Figure 3-18 of the *Knowledge Base*.
   3. Record the parts cost.

   Information:
   Labor: 1 hour
   Parts: 6 spark plugs 135A

   System of Measurement: Time and money

   Compute:
   $1 \times \$35.00 = \$35.00$
   $6 \times \$3.50 = \$21.00$

   Communicate:
   The "Labor" and "Part" sections of the work order look like this.

WORK ORDER

QTY.	PART	PRICE	SERVICE/LABOR	PRICE
6	Spark plugs 135A	$21.00	Tune-up	$35.00
	TOTAL PARTS		TOTAL LABOR	

7. **To Do:** Fill in the costs on the work order.

   Steps:
   1. Find and record the labor cost.
   2. Find the parts cost in Figure 3-18 of the *Knowledge Base*.
   3. Record the parts cost.

   Information:
   Labor: ½ hour
   Parts: 4 qt. oil; oil filter 12J

   System of Measurement: Time and money

   Compute:
   $\frac{1}{2} \times \$35.00 = 0.5 \times \$35.00 = \$17.50$
   $4 \times \$1.76 = \$7.04$
   $1 \times \$6.15 = \$6.15$

   Communicate:
   The "Labor" and "Part" sections of the work order look like this.

WORK ORDER

QTY.	PART	PRICE	SERVICE/LABOR	PRICE
1	Oil filter 12J	$6.15	Oil change	$17.50
4	Qt. oil	$7.04		
	TOTAL PARTS		TOTAL LABOR	

152

8. **To Do:** Fill in the costs on the work order.

Steps:
1. Find and record the labor cost.
2. Find the parts cost in Figure 3-18 of the *Knowledge Base*.
3. Record the parts cost.

Information:
Labor: ½ hour
Part: battery UB-111

System of Measurement: Time and money

Compute:
½ × $35.00 = 0.5 × $35.00 = $17.50
1 × $52.00 = $52.00

Communicate:
The "Labor" and "Part" sections of the work order look like this.

WORK ORDER

QTY.	PART	PRICE	SERVICE/LABOR	PRICE
1	Battery UB-111	$52.00	Check battery Replace battery	n/c $17.50
		TOTAL PARTS		TOTAL LABOR

9. **To Do:** Fill in the costs on the work order.

Steps:
1. Find and record the labor cost.
2. Find the parts cost in Figure 3-18 of the *Knowledge Base*.
3. Record the parts cost.

Information:
Labor: 2½ hours
Part: water pump 328Q

System of Measurement: Time and money

Compute:
2½ × $35.00 = 2.5 × $35.00 = $87.50
1 × $62.27 = $62.27

Communicate:
The "Labor" and "Part" sections of the work order look like this.

WORK ORDER

QTY.	PART	PRICE	SERVICE/LABOR	PRICE
1	Water pump 328Q	$62.27	Replace water pump	$87.50
		TOTAL PARTS		TOTAL LABOR

10. **To Do:** Fill in the costs on the work order.

Steps:
1. Find and record the labor cost.
2. Find the parts cost in Figure 3-18 of the *Knowledge Base*.
3. Record the parts cost.

Information:
Labor: 1 hour

System of Measurement: Time and money

Compute:
1 × $35.00 = $35.00

Communicate:
The "Labor" and "Part" sections of the work order look like this.

WORK ORDER

QTY.	PART	PRICE	SERVICE/LABOR	PRICE
			Adjust brakes	$35.00
		TOTAL PARTS		TOTAL LABOR

TASK 3

11. **To Do:** Complete the final bill.

Steps:
1. Check the labor cost.
2. Write the total labor.
3. Check the parts cost. Write the name and cost of any new parts on the work order.
4. Add the cost of all the parts.
5. Find the total sale.

Information:
Labor and parts were as estimated on the work order.

System of Measurement: Time and money

Compute:
Labor costs: $35.00
Parts: $21.00
Total sale: $35.00 + $21.00 = $56.00

Communicate:
The completed work order and final bill looks like this.

WORK ORDER

DATE: 1/10/9X

NAME: Alan Billings
ADDRESS: 100 Plymouth Ave.
CITY/ZIP: Northtown, IL 50050

PHONE: 555-3498

SERVICE/LABOR	PRICE
Tune-up	$35.00

YEAR	MAKE	MODEL	LICENSE			
90	Ford	Taurus	4RD 111		TOTAL LABOR	$35.00

QTY.	PART		PRICE		
6	Spark plugs 135A		$21.00	TOTAL LABOR	$35.00
				TOTAL PARTS	$21.00
				TOTAL SALE	$56.00
	TOTAL PARTS		$21.00	CUSTOMER SIGNATURE: _____	

---

12. **To Do:** Complete the final bill.

Steps:
1. Check the labor cost.
2. Write the total labor.
3. Check the parts cost. Write the name and cost of any new parts on the work order.
4. Add the cost of all the parts.
5. Find the total sale.

Information:
Labor was as estimated on the work order. An extra part, air filter A26, was used.

System of Measurement: Time and money

Compute:
Labor costs: $17.50
Part: A26 = $12.50.
Total parts: $6.15 + $7.04 + $12.50 = $25.69
Total sale = $17.50 + $25.69 = $43.19

Communicate:
The completed work order and final bill looks like this.

WORK ORDER

DATE: 1/10/9X

NAME: Alice Anderson
ADDRESS: 1 Crescent Blvd.
CITY/ZIP: Northtown, IL 50050

PHONE: 555-1111

SERVICE/LABOR	PRICE
Oil change	$17.50

YEAR	MAKE	MODEL	LICENSE			
82	Chevy	Caprice	5TY 123		TOTAL LABOR	$17.50

QTY.	PART		PRICE		
1	Oil filter 12J		$6.15	TOTAL LABOR	$17.50
4	Qt. oil		$7.04	TOTAL PARTS	$25.69
1	Air filter A26		$12.50	TOTAL SALE	$43.19
	TOTAL PARTS		$25.69	CUSTOMER SIGNATURE: _____	

13. **To Do:** Complete the final bill.

Steps:
1. Check the labor cost.
2. Write the total labor.
3. Check the parts cost. Write the name and cost of any new parts on the work order.
4. Add the cost of all the parts.
5. Find the total sale.

Information:
Labor was as estimated on the work order. Extra parts, 4 spark plugs 172F, were used.

System of Measurement: Time and money

Compute:
Labor costs: $17.50
Parts: 4 of 172F = $4 \times \$2.80 = \$11.20$
Total parts: $\$52.00 + \$11.20 = \$63.20$
Total sale: $\$17.50 + \$63.20 = \$80.70$

Communicate:
The completed work order and final bill looks like this.

WORK ORDER

DATE: 1/10/9X

NAME: Kathryn Kirkpatrick
ADDRESS: 34 Old Mill Lane
CITY/ZIP: Northtown, IL 50050

PHONE: 555-3498

YEAR	MAKE	MODEL	LICENSE
87	Honda	Civic	4GT 222

QTY.	PART	PRICE
1	Battery UB-111	$52.00
4	Spark plug 172F	$11.20
	TOTAL PARTS	$63.20

SERVICE/LABOR	PRICE
Check battery	n/c
Replace battery	$17.50
TOTAL LABOR	$17.50

TOTAL LABOR	$17.50
TOTAL PARTS	$63.20
TOTAL SALE	$80.70

CUSTOMER SIGNATURE: _____

---

14. **To Do:** Complete the final bill.

Steps:
1. Check the labor cost.
2. Write the total labor.
3. Check the parts cost. Write the name and cost of any new parts on the work order.
4. Add the cost of all the parts.
5. Find the total sale.

Information:
Labor was as estimated on the work order. An extra part, 1 belt 472F, was used.

System of Measurement: Time and money

Compute:
Labor costs: $87.50
Part: 472F = $13.50
Total parts: $\$62.27 + \$13.50 = \$75.77$
Total sale: $\$87.50 + \$75.77 = \$163.27$

Communicate:
The completed work order and final bill looks like this.

WORK ORDER

DATE: 1/10/9X

NAME: Herb Jones
ADDRESS: 111 West Birch
CITY/ZIP: Northtown, IL 50050

PHONE: 555-7634

YEAR	MAKE	MODEL	LICENSE
79	Dodge	Dart	367 UUY

QTY.	PART	PRICE
1	Water pump 328Q	$62.27
1	Belt 472F	$13.50
	TOTAL PARTS	$75.77

SERVICE/LABOR	PRICE
Replace water pump	$87.50
TOTAL LABOR	$87.50

TOTAL LABOR	$87.50
TOTAL PARTS	$75.77
TOTAL SALE	$163.27

CUSTOMER SIGNATURE: _____

15. **To Do:** Complete the final bill.

Steps:
1. Check the labor cost.
2. Write the total labor.
3. Check the parts cost. Write the name and cost of any new parts on the work order.
4. Add the cost of all the parts.
5. Find the total sale.

Information:
Labor was 1½ hours. Extra parts, 2 brake pads 673, were used.

System of Measurement: Time and money

Compute:
Labor costs: $1.5 \times \$35.00 = \$52.50$
Parts: $2 \times \$36.00 = \$72.00$
Total parts: $72.00
Total sale: $\$52.50 + \$72.00 = \$124.50$

Communicate:
The completed work order and final bill looks like this.

WORK ORDER

DATE: 1/10/9X

NAME: Sam Avino
ADDRESS: 527 East Elm
CITY/ZIP: Northtown, IL 50050

PHONE: 555-1287

SERVICE/LABOR	PRICE
Adjust brakes	$~~$35.00~~$ $52.50

YEAR	MAKE	MODEL	LICENSE
88	Buick	Regal	SAMMY

TOTAL LABOR	$52.50

QTY.	PART	PRICE
2	Brake pad 673	$72.00
	TOTAL PARTS	$72.00

TOTAL LABOR	$52.50
TOTAL PARTS	$72.00
TOTAL SALE	$124.50

CUSTOMER SIGNATURE: _____

P A R T 5 *Answer Key*

JOB SITUATION 1

TASK 1

1. **To Do:** Find out what maintenance the car needs.

Steps:
1. Find the car's mileage.
2. Read down the column to find the services needed.

Information:
Mileage: 7500 miles

System of Measurement: Miles

Compute:
Use the "7.5" (7500 miles) column. Service: change oil.

Communicate:
The work order looks like this.

SERVICE/LABOR	PRICE
change oil	

2. **To Do:** Find out what maintenance the car needs.

Steps:
1. Find the car's mileage.
2. Read down the column to find the services needed.

Information: Mileage: 15,100

System of Measurement: Miles

Compute:
Use the "15" (15,000 miles) column. Service: change oil, replace oil filter, adjust valve lash, inspect/adjust drive belts.

Communicate:
The work order looks like this.

SERVICE/LABOR	PRICE
change oil	
replace oil filter	
adjust valve lash,	
inspect/adjust drive belts	

3. **To Do:** Find out what maintenance the car needs.

   Steps:
   1. Find the car's mileage.
   2. Read down the column to find the services needed.

   Information: Mileage: 89,121 miles

   System of Measurement: Miles

   Compute:
   Use the "90" (90,000 miles) column in second chart. Service: change oil, replace oil filter, inspect/adjust drive belts, replace spark plugs, replace air filter, apply solvent to choke shaft, check choke heat source.

   Communicate:
   The work order looks like this.

SERVICE/LABOR	PRICE
change oil	
replace oil filter	
inspect/adjust drive belts	
replace spark plugs	
replace air filter	
apply solvent to choke shaft	
check choke heat source	

4. **To Do:** Find out what maintenance the car needs.

   Steps:
   1. Find the car's mileage.
   2. Read down the column to find the services needed.

   Information: Mileage: 59,431 miles

   System of Measurement: Miles

   Compute:
   Use the "60" (60,000 miles) column in first chart. Service: change oil, replace oil filter, adjust valve lash, inspect/adjust drive belts, replace spark plugs, replace air filter, apply solvent to choke shaft, check choke heat source, check ignition timing, replace ignition cables, cap & rotor/inspect distributor.

JOB SITUATION 2

TASK 1

1. **To Do:** Convert centimeters to millimeters. Find the correct part.

   Steps:
   1. Find the correct rule.
   2. Look in the table to see how the two units compare.
   3. Do the computation.

Communicate:
The work order looks like this.

SERVICE/LABOR	PRICE
change oil	
replace oil filter	
adjust valve lash	
inspect/adjust drive belts	
replace spark plugs	
replace air filter	
apply solvent to choke shaft	
check choke heat source	
check ignition timing	
replace ignition cables, cap & rotor/inspect distributor	

5. **To Do:** Find out what maintenance the car needs.

   Steps:
   1. Find the car's mileage.
   2. Read down the column to find the services needed.

   Information: Mileage: 53,000 miles

   System of Measurement: Miles

   Compute:
   Use the "52.5" (52,500 miles) column in first chart. Service: change oil, replace fuel filter, replace oxygen sensor, replace PCV valve, check air injection valves, replace canisters, inspect rubber hoses, check and reset idle, inspect/replace gas tank cap.

   Communicate:
   The work order looks like this.

SERVICE/LABOR	PRICE
change oil	
replace fuel filter	
replace oxygen sensor	
replace PCV valve	
check air injection valves	
replace canisters	
inspect rubber hoses	
check and reset idle speed	
inspect/replace gas tank cap	

Information: Need: 1 cm bolt

System of Measurement:
Length/Metric system

Compute:
To convert to a smaller unit, multiply.
   1 cm = 10 mm
   1 cm × 10 = 10 mm

Communicate:

The purchase order looks like this.

Supplier: Ed's Auto Parts			Ship to:	
Address: 555 West 10th St. Northtown, IL 50050			A&P Auto Center 326 Main St. Chester, IA 40092	
Date: 1/15/9X				
Quantity	Description of Part		Unit Price	Extension Price
1 doz	10-mm bolt		$2.90/doz	
			TOTAL:	

2. **To Do:** Convert millimeters to centimeters. Find the correct part.

   Steps:
   1. Find the correct rule.
   2. Look in the table to see how the two units compare.
   3. Do the computation.

   Information: Need: 35-mm washer

System of Measurement:
   Length/Metric system

Compute:
   To convert to a larger unit, divide.
   $$1 \text{ cm} = 10 \text{ mm}$$
   $$35 \text{ mm} \div 10 = 3.5 \text{ cm}$$

Communicate:
   The purchase order looks like this.

Supplier: Ed's Auto Parts			Ship to:	
Address: 555 West 10th St. Northtown, IL 50050			A&P Auto Center 326 Main St. Chester, IA 40092	
Date: 1/15/9X				
Quantity	Description of Part		Unit Price	Extension Price
1 doz	Belt TH-35		$34.00/doz	
2 doz	3.5-cm washer		$ 3.26/doz	
			TOTAL:	

3. **To Do:** Convert millimeters to centimeters. Find the correct part.

   Steps:
   1. Find the correct rule.
   2. Look in the table to see how the two units compare.
   3. Do the computation.

Information: Need: 125-mm bolt

System of Measurement:
   Length/Metric system

Compute:
   To convert to a larger unit, divide.
   $$1 \text{ cm} = 10 \text{ mm}$$
   $$125 \text{ mm} \div 10 = 12.5 \text{ cm}$$

Communicate:
The purchase order looks like this.

Supplier: Ed's Auto Parts		Ship to:	
Address: 555 West 10th St. Northtown, IL 50050		A&P Auto Center 326 Main St. Chester, IA 40092	
Date: 1/15/9X			
Quantity	Description of Part	Unit Price	Extension Price
2 doz 4 doz	Hose 78-t1 12.5-cm bolt	$27.89/doz $ 3.89/doz	
		TOTAL:	

4. **To Do:** Convert meters to centimeters. Find the correct part.

Steps:
1. Find the correct rule.
2. Look in the table to see how the two units compare.
3. Do the computation.

Information: Need: 1.10-m rod

System of Measurement:
Length/Metric system

Compute:
To convert to a smaller unit, multiply.
$$1 \text{ m} = 100 \text{ cm}$$
$1.10 \text{ m} \times 100 = 110 \text{ cm}$ (Move the decimal point two places to the right.)

Communicate:
The purchase order looks like this.

Supplier: Ed's Auto Parts		Ship to:	
Address: 555 West 10th St. Northtown, IL 50050		A&P Auto Center 326 Main St. Chester, IA 40092	
Date: 1/15/9X			
Quantity	Description of Part	Unit Price	Extension Price
2 doz	110 cm rod	$35.22/doz	
		TOTAL:	

5. **To Do:** Convert centimeters to meters. Find the correct part.

Steps:
1. Find the correct rule.
2. Look in the table to see how the two units compare.
3. Do the computation.

Information: Need: 85-cm tubing

System of Measurement:
Length/Metric system

Compute:
To convert to a larger unit, divide.
$$1 \text{ m} = 100 \text{ cm}$$
$85 \text{ cm} \div 100 = 0.85 \text{ m}$

Communicate:

The purchase order looks like this.

Supplier: Ed's Auto Parts			Ship to: A&P Auto Center	
Address: 555 West 10th St. Northtown, IL 50050			326 Main St. Chester, IA 40092	
Date: 1/15/9X				
Quantity	Description of Part		Unit Price	Extension Price
1 doz	Oil filter 89AB		$67.80/doz	
2 doz	Tire valve 8J		$24.00/doz	
2 doz	0.85 m plastic tubing		$ 5.67/doz	
			TOTAL:	

TASK 2

6. **To Do:** Complete the purchase order.

   Steps:
   1. Find and write each extension price.
   2. Find and write the total price.

   Information:

1 doz	10-mm bolt	$2.90/doz

   System of Measurement: Money

   Compute:

   $1 \times \$2.90 = \$2.90$

   Total: $2.90

   Communicate:

   The completed purchase order looks like this.

	Extension Price
	$2.90
TOTAL:	$2.90

7. **To Do:** Complete the purchase order.

   Steps:
   1. Find and write each extension price.
   2. Find and write the total price.

   Information:

1 doz	Belt TH-35	$34.00/doz
2 doz	3.5-cm washer	$ 3.26/doz

   System of Measurement: Money

   Compute:

   $1 \times \$34.00 = \$34.00$
   $2 \times \$3.26 = \$6.52$
   $\$34.00 + \$6.52 = \$40.52$

   Communicate:

   The completed purchase order looks like this.

	Extension Price
	$34.00
	$ 6.52
TOTAL:	$40.52

8. **To Do:** Complete the purchase order.

   Steps:
   1. Find and write each extension price.
   2. Find and write the total price.

   Information:

2 doz	Hose 78-t1	$27.89/doz
4 doz	12.5-cm bolt	$ 3.89/doz

   System of Measurement: Money

   Compute:

   $2 \times \$27.89 = \$55.78$
   $4 \times \$3.89 = \$15.56$
   $\$55.78 + \$15.56 = \$71.34$

160

Communicate:
The completed purchase order looks like this.

	Extension Price
	$55.78
	$15.56
TOTAL:	$71.34

9. **To Do:** Complete the purchase order.
   Steps:
   1. Find and write each extension price.
   2. Find and write the total price.
   Information:

‖ 2 doz | 110-cm rod | $35.22/doz ‖

   System of Measurement: Money
   Compute:
   $2 \times \$35.22 = \$70.44$
   Total = $70.44

   Communicate:
   The completed purchase order looks like this.

	Extension Price
	$70.44
TOTAL:	$70.44

10. **To Do:** Complete the purchase order.

    Steps:
    1. Find and write each extension price.
    2. Find and write the total price.

    Information:

1 doz	Oil filter 89AB	$67.80/doz
2 doz	Tire valve 8J	$24.00/doz
2 doz	0.85 mm plastic tubing	$ 5.67/doz

    System of Measurement: Money

    Compute:
$$1 \times \$67.80 = \$67.80$$
$$2 \times \$24.00 = \$48.00$$
$$2 \times \$5.67 = \$11.34$$
$$\$67.90 + \$48.00 + \$11.34 = \$127.14$$

    Communicate:
    The completed purchase order looks like this.

	Extension Price
	$67.80
	$48.00
	$11.34
TOTAL:	$127.14

JOB SITUATION 3

TASK 1

1. **To Do:** Read the gauge and write the measurement.

   Steps:
   1. Read the number on the hub.
   2. Read the lines on the hub.
   3. Read the thimble.
   4. Add the measurements.

   Information:
   The hub shows "2" and "0" markings. The thimble shows "0."

   System of Measurement: Inches

   Compute:
   Hub number: 0.2 in
   Hub lines: $0 \times 0.025 = 0$
   Thimble: 0 in
   $0.2 + 0 + 0 = 0.2$ in

   Communicate: 0.2 in

2. **To Do:** Read the gauge and write the measurement.
   Steps:
   1. Read the number on the hub.
   2. Read the lines on the hub.
   3. Read the thimble.
   4. Add the measurements.

Information:
    The hub shows "2" and "0" markings. The thimble shows "10."

System of Measurement: Inches

Compute:
    Hub number: 0.2 in
    Hub lines: $0 \times 0.025 = 0$
    Thimble: 0.010 in
    $0.2 + 0.010 = 0.210$ in

Communicate: 0.21 in

3. **To Do:** Read the gauge and write the measurement.

    Steps:
        1. Read the number on the hub.
        2. Read the lines on the hub.
        3. Read the thimble.
        4. Add the measurements.

    Information:
        The hub shows "2" and "2" markings. The thimble shows "1."

    System of Measurement: Inches

    Compute:
        Hub number: 0.2 in
        Hub lines: $2 \times 0.025 = 0.050$ in
        Thimble: 0.001 in
        $0.2 + 0.050 + 0.001 = 0.251$

    Communicate: 0.251 in.

4. **To Do:** Read the gauge and write the measurement.

    Steps:
        1. Read the number on the hub.
        2. Read the lines on the hub.
        3. Read the thimble.
        4. Add the measurements.

Information:
    The hub shows "3" and "1" markings. The thimble shows "0."

System of Measurement: Inches

Compute:
    Hub number: 0.3 in
    Hub lines: $1 \times 0.025 = 0.025$ in
    Thimble: 0.000 in
    $0.3 + 0.025 + 0 = 0.325$ in

Communicate: 0.325 in

5. **To Do:** Read the gauge and write the measurement.

    Steps:
        1. Read the number on the hub.
        2. Read the lines on the hub.
        3. Read the thimble.
        4. Add the measurements.

    Information:
        The hub shows "3" and "3" markings. The thimble shows "20."

    System of Measurement: Inches

    Compute:
        Hub number: 0.3 in
        Hub lines: $3 \times 0.025 = 0.075$ in
        Thimble: 0.020 in
        $0.3 + 0.075 + 0.020 = 0.395$ in

    Communicate: 0.395 in

PART 6 *Answer Key*

JOB SITUATION 1

TASK 1

1. **To Do:** Draw a graph to show sales for each brand.

    Steps:
        1. Create a table for the data.
        2. Choose a label for each axis.
        3. Create a scale for the numerical axis.
        4. Write the information on the other axis.
        5. Draw a bar for each pair of numbers in your table.
        6. Title the graph.

Information:
    Goodyear: 275 tires   Michelin: 100 tires
    General: 200 tires

System of Measurement: Tires

Compute:
    1. Table:

Brand	Number of Tires Sold
Goodyear	275
Michelin	100
General	200

2. Labels: "Tires Sold," "Brands of Tires"

3. Scale: 0 to 300 Interval: 25

4. Bars: See the graph below.

5. Title: "Sales of Brands of Tires"

Communicate:

The graph looks like this.

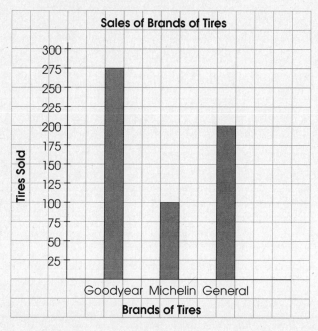

2. **To Do:** Draw a graph to show sales for each brand.

Steps:

1. Create a table for the data.

2. Choose a label for each axis.

3. Create a scale for the numerical axis.

4. Write the information on the other axis.

5. Draw a bar for each pair of numbers in your table.

6. Title the graph.

Information:

Goodyear: 375 tires   Michelin: 250 tires
General: 325 tires

System of Measurement: Tires

Compute:

1. Table:

Brand	Number of Tires Sold
Goodyear	375
Michelin	250
General	325

2. Labels: "Tires Sold," "Brands of Tires"

3. Scale: 0 to 400 Interval: 25

4. Bars: See the graph below.

5. Title: "Sales of Brands of Tires"

Communicate:

The graph looks like this.

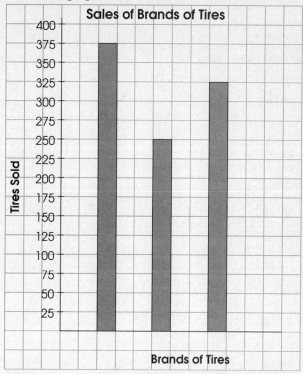

TASK 2

3. **To Do:** Draw a line graph to show tire sales each month.

Steps:

1. Create a table for the data.

2. Choose a label for each axis.

3. Create a scale for the numerical axis.

4. Write the information on the other axis.

5. Draw a dot for each pair of numbers in your table. Connect the dots.

6. Title the graph.

Information:

Sales were: January—45, February—35, March—40, April—45, May—60, June—65.

System of Measurement: Tires

Compute:

1. Table:

Month	Tire Sales
Jan.	45
Feb.	35
Mar.	40
Apr.	45
May	60
June	65

2. Labels: "Tires Sold," "Brands of Tires"

3. Scale: 0 to 70 Interval: 5

4. Lines: See the graph below.

5. Title: "Tire Sales Jan. to June"

Communicate:
The graph looks like this.

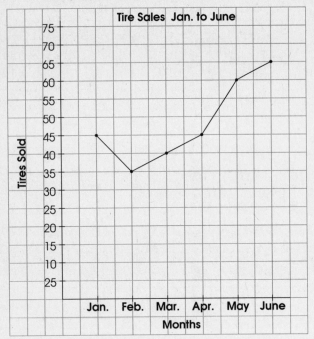

Communicate:
The graph looks like this.

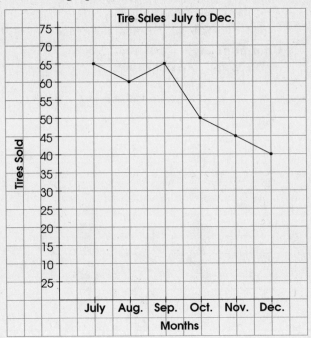

4. **To Do:** Draw a line graph to show tire sales each month.

   Steps:
   1. Create a table for the data.
   2. Choose a label for each axis.
   3. Create a scale for the numerical axis.
   4. Write the information on the other axis.
   5. Draw a dot for each pair of numbers in your table. Connect the dots.
   6. Title the graph.

   Information:
   Sales were: July—65, August—60, September 65, October—50, November—45, December—40.

   System of Measurement: Tires

   Compute:
   1. Table:

Month	Tire Sales
July	65
Aug.	60
Sept.	65
Oct.	50
Nov.	45
Dec.	40

   2. Labels: "Tires Sold," "Brands of Tires"
   3. Scale: 0 to 70 Interval: 5
   4. Lines: See the graph below.
   5. Title: "Tire Sales July to December"

TASK 3

5. **To Do:** Make a circle graph to show why people bought from A&P Auto.

   Steps:
   1. Create a table for the data.
   2. Find the percent of the whole for each item.
   3. Find the number of degrees for each item.
   4. Draw a circle and draw an angle for each item of data. Label it.
   5. Title the graph.

   Information: Total of 1000 people.
   Reasons: Prices—500 people, Brands—200 people, Location—300

   System of Measurement: People

   Compute:
   1. Table:

Reason for Buying at A&P Auto	Number of Customers Giving this Reason
Prices	500
Brands	200
Location	300

   2. Percentages:
   "Prices" chosen by 500 out of 1000 people. What percent is 500 of 1000?

$$n \times 1000 = 500$$
$$\div 1000 \qquad \div 1000$$
$$n = 0.50 = 50\%$$

"Brands" chosen by 200 out of 1000. What percent is 200 of 1000?

$n \times 1000 = 200$

$n = 0.2 = 20\%$

"Location" chosen by 300 out of 1000. What percent is 300 of 1000?

$n \times 1000 = 300$

$n = 0.3 = 30\%$

3. Angles:

"Prices" is 50%. The angle is 50% of 360°. $360° \times 0.50 = 180°$

"Brands" is 20%. Angle is 20% of 360°. $360° \times 0.20 = 72°$

"Location" is 30%. Angle is 30% of 360°. $360° \times 0.30 = 108°$

4. Draw the circle and wedges.

5. Title: "Reasons for Buying at A&P Auto"

Communicate:

The graph looks like this.

Reasons for Buying at A&P Auto

6. **To Do:** Make a circle graph to show how satisfied people are with the service at A&P Auto.

Steps:

1. Create a table for the data.

2. Find the percent of the whole for each item.

3. Find the number of degrees for each item.

4. Draw a circle and draw an angle for each item of data. Label it.

5. Title the graph.

Information: Total of 1000 people.
Very Satisfied—200 people, Satisfied—700 people, Not Satisfied—100 people

System of Measurement: People

Compute:

1. Table:

How Satisfied Are People with A&P Auto?	Number of Customers Giving this Reason
Very Satisfied	200
Satisfied	700
Not Satisfied	100

2. Percentages:
Very Satisfied: 20%
Satisfied: 70%
Not Satisfied: 10%

3. Angles:
Very Satisfied: 72°
Satisfied: 252°
Not Satisfied: 36°

4. Draw the circle and wedges.

5. Title: "Degree of Satisfaction"

Communicate:

The graph looks like this.

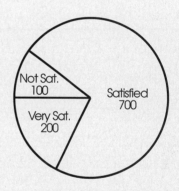

Degree of Satisfaction